# Accessible Algebra

# Accessible Algebra

30 Modules to Promote Algebraic Reasoning,
Grades 7–10

Anne M. Collins
and
Steven R. Benson

www.stenhouse.com

Portland, Maine

Stenhouse Publishers
www.stenhouse.com



Library of Congress Cataloging-in-Publication Data

Names: Collins, Anne, 1950- | Benson, Steven 1961- Accessible algebra: 30 modules to promote algebraic reasoning, grades 7–10 / Anne M. Collins and Steven R. Benson. Description: Portland, Maine : Stenhouse Publishers, [2017]

Identifiers: LCCN 2016040686 (print) | LCCN 2016040854 (ebook) | ISBN 9781625310668 (pbk.: alk. paper) | ISBN 9781625310675 (ebook)

Subjects: LCSH: Algebra--Study and teaching (Middle school) | Algebra--Study and teaching (Secondary)

Classification: LCC QA159 .C635 2017 (print) | LCC QA159 (ebook) | DDC 512.9--dc23

LC record available at https://lccn.loc.gov/2016040686

Cover design, interior design, and typesetting by Victory Productions Inc.

Manufactured in the United States of America

PRINTED ON 30% PCW
RECYCLED PAPER

23 22 21 20 19 18 17    9 8 7 6 5 4 3 2 1

This book is dedicated to the staff of the Center for Mathematics Achievement and the faculty of the Division of Natural Sciences and Mathematics who helped us through this process.

# Contents

# Acknowledgments

We are deeply indebted to the many teachers who welcomed us into their classrooms and gave us permission to observe their teaching and their students learning. A special thanks to the wonderful teachers in the White Brook Middle School, Easthampton, Massachusetts, who even allowed us to teach some lessons to ensure that what we are proposing teachers do actually does work and that students benefit from the activities. Thanks also to the teacher participants in the Massachusetts Mathematics and Science Partnership (MMSP) program in Springfield, Brockton, Quincy, and Weymouth, Massachusetts, for field-testing many of our problems and for donating reams of student work to us and our project. A special thanks to Catherine Hill from Haverhill, Massachusetts, who continues to send us pictures of her students enjoying many of the activities from this book.

We are extremely appreciative of the time Dr. Roser Gine and Stephen Yurek spent working through the problems and solutions contained within this book.

We also want to express our sincere appreciation to Cindy Orrell for her work editing the manuscript. A very special thanks also to Toby Gordon at Stenhouse for her trust that this work would be completed. We believe this publication will provide algebra teachers with a rich set of lessons ready for implementation in their classrooms.

# Introduction

Most people realize they need to know arithmetic, but far fewer understand why they should know algebra. The need to know algebra is not as obvious as the need to be proficient in arithmetic. Yet, algebra is the language of the generalization of arithmetic. Should you need to do something once, then arithmetic usually suffices; however, if you need to do something repetitively, then you need algebra as a tool for generalizing the steps you do arithmetically, regardless of the numbers involved.

- Algebra is the language through which patterns are described.
- Algebra is the language used to describe relationships between quantities.
- Algebra is the language that is used to solve certain types of arithmetic problems.
- Algebra helps people develop new ways of thinking.

Robert Moses, founder of the Algebra Project, and others contend that algebra is a gatekeeper course and that the right to take algebra is a civil rights issue. Too often underserved students—including those living in poverty, those with special needs, and many immigrant, emergent bilingual, and black students—are excluded from the grade eight algebra course, yet when given an experiential approach to algebra, these students are able to demonstrate proficiency. The Algebra Project engaged students in doing algebra. Grade eight students went for rides on the Massachusetts Bay Transit Authority in Boston to model positive and negative integers and were able to demonstrate a greater understanding of the relevance of algebra after all their experiences. We agree with this approach to teaching algebra but recognize its constraints, so we wrote this book, which is filled with activities, modeling, and problem solving that can be done in classrooms.

We hope that this book will be used by teachers of prealgebra or algebraic concepts and/or a traditional algebra course, as well as when introducing functions. We include resources, materials, problems, and games designed to be engaging for students and aligned with the Expressions and Equations, Functions, and High School Algebra domains.

## How to Use This Book

Each of the thirty lessons in this book identifies the focal domain and standard(s) for each lesson. Each lesson includes common misconceptions and challenges that many students face together with suggestions for how a teacher might prevent those misconceptions from developing. The main lesson itself provides examples of how different teachers interact with students in their classrooms as they guide their students toward meeting their learning goals. These classroom scenarios provide suggestions to teachers for how they might assign a particular problem or activity, how to include formative assessment strategies, and how to group students. This section also provides fertile material for discussion within a mathematics department meeting or in professional learning communities. Discussing how and why student misconceptions arise is important as teachers strive to avoid partial understandings and erroneous ideas about concepts.

The "In the Classroom" sections are either representations of what happened in one specific classroom or a combination of effective teaching seen in multiple classrooms. The quotes from students are real. We have met and listened to extraordinary students explaining how they are thinking about a problem or what problem-solving strategies they use. We have observed students struggle to pose a question that might help them understand more deeply the mathematics they are investigating, and we have witnessed high energy in many classrooms while students are working through the activities. Some modules include tasks or problems that might be used in activity centers, for individualized challenges or remediation, or for homework. The center activities are designed to allow students to work at their own level, to challenge students in need of a challenge, or to reinforce previously learned concepts. The "In the Classroom" section also suggests instructional strategies, activities, and samples of student thinking, including connections to the eight Standards for Mathematical Practice (SMPs).

The section on "Meeting Individual Needs" suggests some strategies for students who are struggling or who need a greater challenge. The suggestions in this section include some remediation strategies or further activities to challenge those students who need more in-depth work and some very rigorous problems for students who like to go deeper into a concept than what is typically possible in a heterogeneous classroom.

The final section of each lesson is the "Additional Readings/Resources." We understand how busy teachers are and that finding good research articles or classroom vignettes can be very time-consuming, so we included some references for each lesson. We also suggest that all teachers visit the National Council of Teachers of Mathematics website, which has a plethora of wonderful resources, interactive activities, and readings.

We hope you agree that by embracing the old Chinese proverb, "Tell me, and I will forget. Show me, and I will remember. Involve me, and I will understand," all our students will have access and opportunity to enjoy a rich and fulfilling experience while studying algebra.

We encourage you to keep this resource on your desk or next to your plan book so that you will have these ideas at your fingertips throughout the year. Of course, no single book can contain all of the mathematics pertaining to any given topic. For more detailed discussions about mathematical and teaching issues raised in this book (and elsewhere), visit our blog *Accessible Algebra* at stenhouse.com/accessiblealgebra.

## Additional Reading/Resources

Collins, Anne, and Linda Dacey. 2010. *Zeroing in on Number and Operations: Key Ideas and Common Misconceptions.* Portland, ME: Stenhouse.

———.2011. *The Xs and Whys of Algebra: Key Ideas and Common Misconceptions.* Portland, ME: Stenhouse.

———.2013. *It's All Relative: Key Ideas and Common Misconceptions.* Portland, ME: Stenhouse.

National Council of Teachers of Mathematics (NCTM). 2000. *Principles and Standards for School Mathematics*. Reston, VA: NCTM.

National Council of Teachers of Mathematics (NCTM), National Governors Association (NG), and Council of Chive State School Officers (CCSSO). 2010. *Reaching Higher: The Common Core State Standards Validation Committee: A Report from the National Governors Association Center for Best Practices and the Council of Chief State School Officers*. Washington, DC: NGA Center and CCSSO.

# Expressions

## 1. Order of Operations

**DOMAIN:** **Expressions and Equations**

**STANDARDS:** **6.EE.2a.** Evaluate expressions at specific values of their variables. Include expressions that arise from formulas used in real-world problems. Perform arithmetic operations, including those involving whole number exponents, in the conventional order when there are no parentheses to specify a particular order (Order of Operations).

**6.EE.3.** Apply the properties of operations to generate equivalent expressions.

**7.EE.1.** Apply properties of operations as strategies to add, subtract, factor, and expand linear expressions with rational coefficients.

### Potential Challenges and Misconceptions

Many students struggle with simplifying expressions with several operations because they do not understand that they are actually writing equivalent expressions as they conduct each operation. Too often students are asked to memorize the mnemonic device PEMDAS without understanding it or what its limitations are. Many find that using PEMDAS causes them to incorrectly simplify expressions, because the mnemonic neglects to include *multiply OR divide from LEFT to RIGHT* and *add OR subtract from LEFT to RIGHT*. It is most helpful to provide students with a template that helps them organize their work systematically.

### In the Classroom

Parentheses
Exponents
OR
Multiplication | Division
OR
Addition | Subtraction

To help students visualize the order in which expressions are simplified, one teacher uses a "hopscotch" diagram. He draws three large classroom-sized diagrams on the floor with liquid shoe polish and labels the operations as shown. He also provides his students with heavyweight plain paper and tells them to make their own diagram to refer to as they write equivalent expressions.

Once all students have the hopscotch diagram at their tables, he projects an expression on the board. He invites three student volunteers to use the floor diagram, moving to the appropriate steps and writing the equivalent expressions. (Notice, the term *simplify the expression* is not being used. This is intentional since every step of the "simplification process" is writing equivalent expressions.) Each student volunteer records the steps and resulting mathematical expression after each operation is conducted. While the three students are working on the floor diagrams, the rest of the class is going through the same process on their desk-sized representations.

After an allotted period of time, this teacher asks the three volunteers for the final expression and records it on the board. He then asks the class if anyone got a different answer or if they did it differently. He quickly asks a volunteer to describe the order in which they completed the steps used. He repeats this activity until every student has an opportunity to use the floor-sized diagrams, and as the activity continues, he makes the expressions more and more complicated. See *Classroom Expressions* on page A1-2 in the appendix.

Next, this teacher groups his students into triads and assigns the number 1, 2, or 3 to each student. He invites the number 1s to the board to record an expression that he dictates. He instructs these students to complete the first computation and write an equivalent expression. These students then sit down, and the number 2s go to the board to do the next step. This continues until the equivalent expression cannot be simplified further. While students rotate turns at the board, they are also completing the expression at their seats so that every student completes every expression. This teacher repeats this process, dictating expressions and inviting students to work on writing equivalent expressions until he's confident that students understand the process. Next, he assigns the reproducible *In What Order?* on page A1-3 of the appendix.

Students in grade seven were asked to write equivalent expressions for the expression $^{-}4 - 3(12 - 5) \div 3 \cdot {}^{-}2^2$ (item 4 from *In What Order?*). Classroom teachers predicted the greatest difficulty would be computing $^{-}2^2$, which did give many students difficulty but, surprisingly, not as many students as expected. Here are samples of student work showing some of the challenges and misconceptions.

### Student A

Notice that Student A attempts to do each calculation above the original expression. This lacks the structure and organization that is helpful to most students: writing each equivalent expression underneath the previous one. This student inaccurately subtracts 12 – 5 to get 6 instead of 7 but, more importantly, subtracts before multiplying the quantity of 12 – 5 by 3. This student also adds an equal sign to indicate the solution, perhaps thinking that he or she is solving an equation rather than writing a series of equivalent expressions or "simplifying an expression."

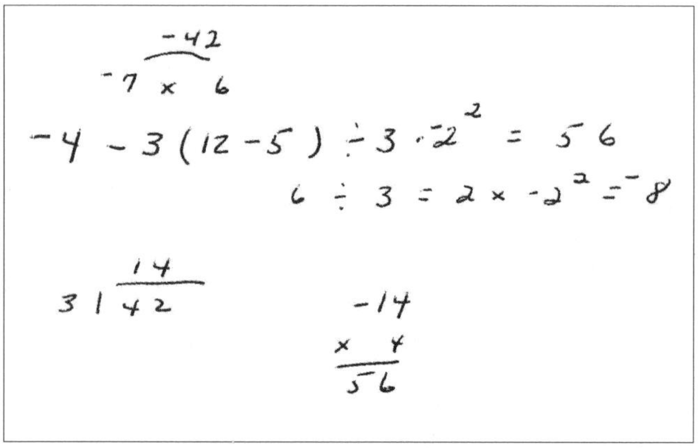

This teacher gives students the choice of using lined paper or the *In What Order? Graphic Organizer* on page A1-4 in the appendix. He cautions students who choose lined paper to record each expression on a separate line and directly below the one before. This provides the teacher a more logical and sequential way of understanding his students' thinking.

### *Student B*

$-4 - 3(12 - 5) \div 3 * -2^2$

This student correctly follows the order of operations but makes errors calculating with negative numbers. When simplifying $^{-}2^2$, he or she makes the classic error of raising $^{-}2$ to the second power instead of raising 2 to the second power then multiplying by $^{-}1$. This student also computes $^{-}4 - 28$ and gets a $^{+}24$.

### *Student C*

$-4 - 3(12 - 5) \div 3 * -2^2$

This student computes using the correct order of operations and accurately simplifies $^{-}2^2$. Unfortunately, when the student moves on to the fourth step, he or she neglects the negative sign: instead of multiplying $^{-}7$ by $^{-}4$, the student multiplies $^{-}7$ by $^{+}4$. This student also inappropriately adds an equal sign to the expression.

As a result of seeing the students' work, this teacher began using the *In What Order? Graphic Organizer* to help his students work systematically when using the order of operations.

Periodically, this teacher assigns the order of operations activity, *Equivalent Expressions*, in various activity centers to allow his students to practice and hone their mathematical skills. *Equivalent Expressions* can be found on page A1-6 in the appendix. Note that teachers should cut out the cards and group them as described on the reproducible before giving them to students. Students select the equivalent expressions strips and glue them in the appropriate order in their notebooks.

## Meeting Individual Needs

Some students benefit from using a graphic organizer that lists the operations in the order they should be calculated, line by line. This not only helps them choose the appropriate operation but also helps them organize and structure the results of forming equivalent expressions. (See *In What Order? Graphic Organizer* in the appendix.) Assign the reproducible *In What Order?*, together with the graphic organizer, to help students internalize the hierarchy represented by order of operations.

## Additional Reading/Resources

Blackwell, Sarah B. 2003. "Operation Central: An Original Play Teaching Mathematical Order of Operations." *Teaching Mathematics in the Middle School* 9 (1): 5.

# 2. Writing Expressions from Tables

**DOMAIN: Expressions and Equations**

**STANDARD: 6.EE.6.** Use variables to represent numbers and write expressions when solving a real-world or mathematical problem; understand that a variable can represent an unknown number, or, depending on the purpose at hand, any number in a specified set.

## Potential Challenges and Misconceptions

Writing an expression from a table presents many challenges for students. One challenge is in understanding which values to focus on. A student may look for the relationship between each input and its output or look for a recursive relationship in the output values, that is, how each subsequent value in the output is related to the output that precedes it. Looking for the recursive relationship in the outputs is common, and while it can be a step to finding the relationship between inputs and outputs, it has its limitations.

Another challenge for students is when the input values do not increase consecutively. When the input values increase by values that are not consecutive, students must look at the ratio between the input and the output, something most students tend not to do. In addition, it is not possible to write a recursive rule if the input does not grow consecutively; students often give up rather than shift focus and look for the relationship between each input and its output. To overcome these challenges, students need multiple opportunities to work with a variety of tables in which the input grows in a variety of ways, for example, inputs placed in the table randomly so that students must rewrite the table in numerical order, skip counting in the input so students must look at the ratio between the output and the input, and tables in which the input grows consecutively.

## In the Classroom

In one classroom, this teacher begins by presenting this table and a range question designed to let the teacher understand what students know about a particular concept, what they misremember, or what misconceptions they bring to the concept. She asks, "Using words, what rule can you write about the data in this table?" Notice that the input data are not presented in numerical order. The teacher does this intentionally in order to guide students to examine the input as well as the output rather than jump to the conclusion that the input values are in numerical order.

| **Input** | 12 | 5 | 14 | 35 | 21 | 39 | 36 | 22 |
|---|---|---|---|---|---|---|---|---|
| **Output** | 3 | 5 | 7 | 7 | 7 | 13 | 3 | 11 |

This teacher has previously introduced the idea of translating information from a table or a problem into a verbal expression before attempting to write a symbolic expression. Using this technique, she has found that all students demonstrate some level of proficiency, while when they are asked to write only a symbolic expression, those same students who tend to struggle continue that struggle or even quit trying. This teacher sits with each group for a short period of time observing, listening, identifying strategies students are using, and thinking about critical questions to ask her students.

After an allotted period of time, this teacher asks student volunteers what they are thinking. Many students simply state that they don't know, but Alex states, "I organized the columns so that the inputs were in numerical order, so I had 5, 12, 14, 21, 22, 35, 36, and 39. Then I saw that the outputs are the greatest prime factor of the input. This was tricky, because usually tables are arranged in numerical order." She pauses. "I don't know how to write a mathematical expression for greatest prime factor though."

This teacher asks her students to brainstorm in small groups how they might write this expression. She listens in and recognizes that her students do not have the expertise to do this. She suggests that they agree on a subscript that might help. She suggests the students brainstorm together how they might represent this and suggests that for the number 15, she might record: $15_{(gpf)} = 5$. She also directs them to graph the greatest prime factors (GPFs) for the positive integers 1 to 30. She tells her students to work in pairs and make a list of observations for this graph. (See the GPF graph on page A1-8 in the appendix.)

Next, this teacher asks her students to write a verbal expression followed by a mathematical expression for each of these tables. Note that in each table, the inputs are arranged in numerical order, but the difference between one input and the next is not the same. She tells students to use the last column, with input $n$, to write the mathematical expression.

| **Input** | 1 | 3 | 6 | 7 | 10 | $n$ |
|---|---|---|---|---|---|---|
| **Output** | 1 | 9 | 36 | 49 | 100 | |

She provides a short time for her students to write the expressions before asking volunteers to share their thinking. Recall that while her students are working, this teacher walks around the room, observing strategies, listening to explanations and excuses, and thinking about who should report out first.

Laila reports, "The verbal expression is, 'Take the input and multiply it by itself.' So my mathematical expression is $n \times n$."

Dominic says, "I got almost the same thing, but I said it is the number in the input squared, so for $n$, the mathematical expression would be $n^2$."

This teacher next presents the following two tables and challenges her students to write both the verbal and symbolic expressions for them.

| **Input** | 1 | 4 | 5 | 7 | 11 | 13 | 15 | $n$ |
|---|---|---|---|---|---|---|---|---|
| **Output** | 2 | 11 | 14 | 20 | 32 | 38 | 44 | |

| **Input** | 3 | 4 | 5 | 6 | 7 | 8 | 9 | $n$ |
|---|---|---|---|---|---|---|---|---|
| **Output** | 14 | 18 | 22 | 26 | 30 | 34 | 38 | |

After an allotted period of time, she invites student volunteers to go to the document camera and share their work. Dimitri volunteers and presents his work for the first table.

"In the first table, the inputs are in order, but there are missing numbers." He points to inputs 1 and 4. "See—2 and 3 are missing, and 6 is missing between 5 and 7. So I had to do a guess-and-check with one pair. I picked 5 because 5s are easy. The input is 5 and the output is 14. I know 14 is almost 15, which is 5 × 3—just 1 less, actually. So that was my guess: multiply by 3 and subtract 1. I checked that with input 4 and it worked: 4 times 3 is 12, minus 1 is 11. I checked one more, just in case: 1 times 3 is 3, minus 1 is 2—and it worked, too. So my mathematical expression is $3n - 1$."

multiply the input by three then subtract one
3i - 1 so n is 3n - 1

This teacher asks whether anyone wrote a different phrase or expression. When no one volunteers, she moves on to the second table. Vince asks to show his work and shares the following.

Vince explains, "Alex had a good idea about checking the inputs. Look, these are all in number order, and there are not any missing numbers. So I looked at the outputs. Each one is 4 more than the one before; when the input goes up by 1, the output goes up by 4. So I figured the rule has to have 'times the input by 4' in it. I checked that with input 3. Three times 4 is 12, and this output is 14, so 'times 4' can't be the rule. Maybe the rule is 'times the input by 4 and add 2.' That works for all of the inputs, so my expression is $4n + 2$."

times the input by 4 then add two
4n + 2

Again this teacher asks whether anyone has anything different. Bonnie asks whether she can ask a question. She poses, "Mine is almost the same, but can you put a number in the verbal expression? I wrote the number *four* out but Vince put in a 4." The class discusses this point with some students agreeing that they should write out the number and not the symbol until they write the symbolic expression. The class agrees that it does not really matter, but they should use the word.

After this unexpected conversation about form, and once this teacher thinks students have gained proficiency with writing expressions from tables, she assigns the *What's My Expression?* reproducible from page A1-9 in the appendix.

## Meeting Individual Differences

Many students struggle translating English word phrases to mathematics and from mathematics to the English word phrases. To help those students who need additional support, this teacher has made a set of cards, *Match the Equivalents*, on page A1-10 in the appendix, with the verbal phrases and the corresponding mathematical expressions. Depending on how many students need this extra reinforcement, she either pairs students to work together or assigns the matching to an individual student who glues the matching pairs together on paper or in a notebook.

## Additional Reading/Resources

Moseley, Bryan. 2004. "Express Yourself with Algebra? This Game Delivers!" *Teaching Mathematics in the Middle School* 10 (1): 38–40.

# 3. Growing Patterns

**DOMAIN:** **Expressions and Equations**

**STANDARDS:** **6.EE.6.** Use variables to represent numbers and write expressions when solving a real-world or mathematical problem; understand that a variable can represent an unknown number, or, depending on the purpose at hand, any number in a specified set.

**7.EE.2.** Understand that rewriting an expression in different forms in a problem context can shed light on the problem and how the quantities in it are related.

## Potential Challenges and Misconceptions

It is often challenging for students to look at a geometric representation of a growing pattern and write an expression that actually describes how the pattern is growing. Often teachers encourage or require students to write all their expressions in factored form, which does not typically illustrate what is happening in the series visually. The expanded form is often more closely related to what is happening as the pattern is growing. When students are able to leave their representations in the form that they believe best represents how the data are growing, they can typically explain what is happening both in the pattern and in their expressions.

## In the Classroom

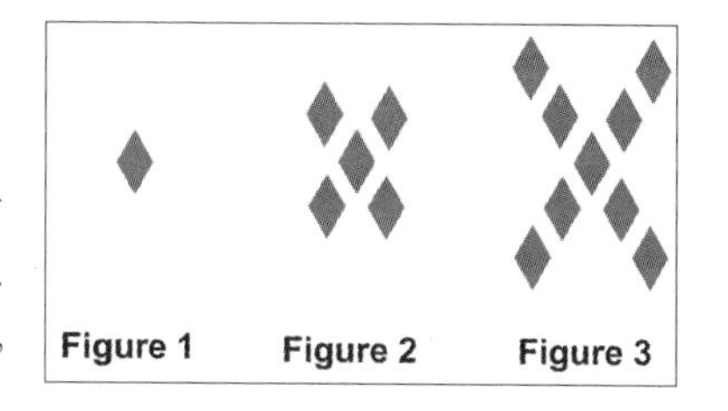

Figure 1 Figure 2 Figure 3

One teacher projects the growing pattern from the reproducible *Growing Figures 1* from page A1-12 of the appendix. She challenges her students to study the figures and think about what they are observing. After an allotted period of time, this teacher asks students to turn to their shoulder partners and share their observations. Once the sharing is completed, this teacher records her students' observations. They include: "I see the first diamond in the middle of all three patterns." "I see the first rhombus is surrounded by four more rhombi in the second example and by eight in the third one." "I see that the number of tails is growing by one in the second and third pictures." "I see two diagonals in the second and third picture." "I see four tails growing by the same amount with a rhombus in the middle."

Once all the students have shared their observations, this teacher suggests her students write a mathematical expression for what the $n$th iteration would look like.

### *Student A*

As the students work, this teacher walks around the room observing how each student is thinking about the problem. She notices that Student A begins with a table and from there tries various expressions, checking to see whether each expression works for each iteration. This student typically tries to ignore using diagrams and pictures and starts with the abstract.

| figure | NUMBER |
|---|---|
| 1 | 1 |
| 2 | 5 |
| 3 | 9 |
| 4 | 13 |
| 5 | 17 |
| 6 | 21 |
| 7 | 25 |
| n | |

+4 +4 +4 +4 +4 +4

~~4n 4(1) = 4~~
4n 4(2) = 8

~~n + 4~~
~~1 + 4 = 4~~

~~2n + 1~~
~~2(2) + 1 = 5~~
~~2(3) + 1 = 7~~

***Student B***

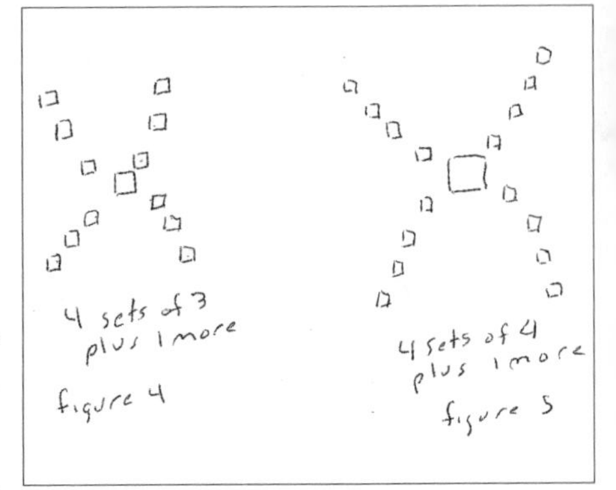


This student begins by sketching what would happen in iterations 4 and 5. Under each drawing, she records what she notices. As she records *4 sets of 3 plus 1 more* and 4 sets *of 4 plus 1 more*, she first writes $4n + 1$, but quickly realizes that it does not work.

This teacher suggests that Students A and B work together to see whether they can combine their thinking to derive the expression. It does not take the two students long to arrive at the correct expression, $4(n - 1) + 1$. This teacher asks these two students to share their original thinking and how they arrived at the expression. This collaboration supports Standard for Mathematical Practice (SMP) 3—construct viable arguments and critique the reasoning of others.

The teacher then distributes the *What's My Rule?* reproducible on page A1-13 of the appendix, which contains this growing pattern:

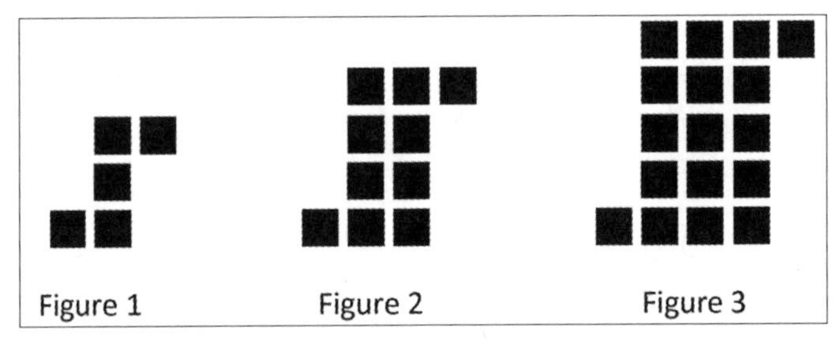


She pairs students, matching a student who typically uses pictures and diagrams with one who relies on the abstract. After an allotted period of time, she asks pairs to share their observations. Annie says, "I see a square in the middle of two rectangles and two extra squares." This teacher asks Annie to go to the board and use a marker to outline the square she is seeing and to use a different color to emphasize the two extra squares she sees.

When Simon says, "I see vertical rectangles plus two extra squares," this teacher asks Simon to go to the board and outline the rectangles that he sees. She prompts her students to examine both Annie's and Simon's outlines and compare them to what they did.

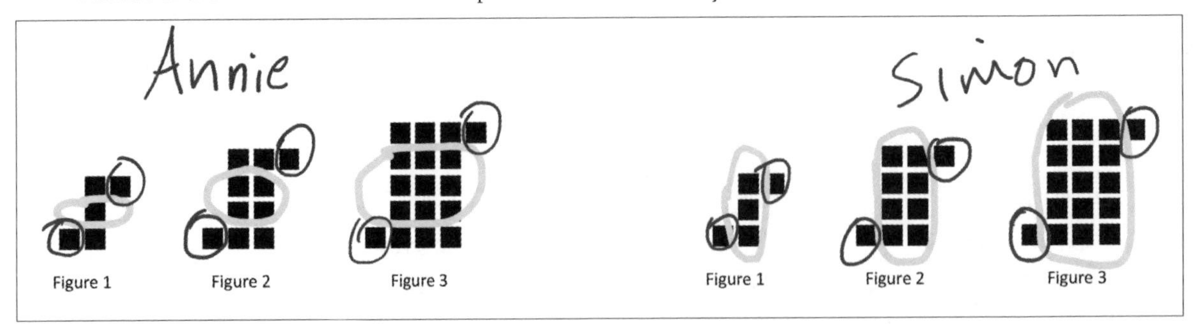


Isaiah excitedly shares the fact that, "I counted the number of squares and made a table and used finite differences and determined the pattern is quadratic." Isaiah has older siblings and always wants to share the shortcuts he learns at home. This teacher thanks Isaiah and states that when they start working with quadratics they will come back to his ideas. She validates his contribution by writing the term *finite differences* on the parking lot section of the board. This is an area where this teacher records topics, vocabulary, or other interesting ideas with the intention of going back to them when appropriate.

After asking whether anyone has different observations, this teacher provides the pairs with centimeter blocks and tells them to build the figures. She hopes the cubes will help her students determine how the figures are changing, which is necessary if they are to be successful in writing the expressions. After an allotted period of time, she returns to her first question and asks her students to describe how they went about building the model of the figures. Most students began with one of two methods: placing the square in the middle and then adding rectangles, followed by the extra two cubes, or building the rectangle at the bottom and then the square in the middle, followed by the rectangle on top.

When it comes time for students to write their expressions, this teacher gives them time to work individually so she might see what help might be necessary. As she observes, she notices that students who build the middle square first followed by the rectangles and finally the additional two blocks have the easiest time writing an algebraic expression. These students write $n \times n + 2 \times n + 2$ and then many follow up with $n^2 + 2(n + 1)$. One student writes $n^2 + 2n + 2$, but when asked to explain where in the figure those values were, the student is unable to make a connection. This teacher records each expression on the conjecture board so that she, and the students, can come back to any expression that is not clear.

Another student says he found a diagonal pattern but isn't sure how to describe it or to translate it into an equation. He says he needs more time to think about it. Before moving on, this teacher asks her students if anyone could explain the relationship between the following two expressions: $n^2 + 2(n + 1)$ and $n^2 + 2n + 2$. Many students raise their hands and Maggie explains, "If you use the distributive property on the first expression, you get the second expression because $2(n + 1)$ is the same as $2n + 2$." After recording the various expressions on the board, this teacher asks her students which of the expressions is the most descriptive and which expressions do not allow them to visualize what the pattern might look like.

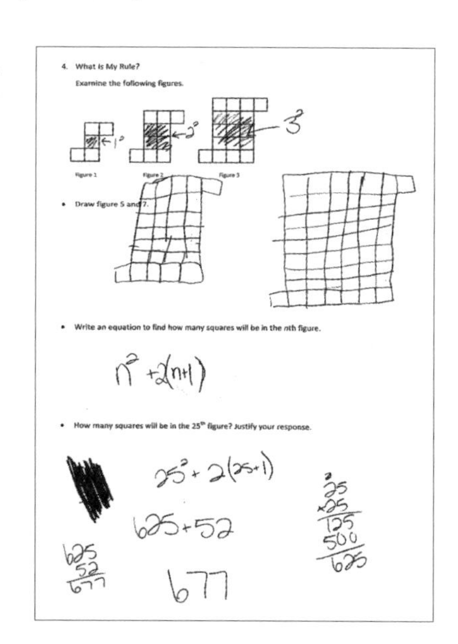


### *Student A*

This work illustrates the fact that this student sees the square in the middle of each figure. Notice that in each figure the square is highlighted. This student is able to write an accurate expression and substitutes into the expression to calculate the value for the twenty-fifth figure in the sequence. When asked how she determined the expression, she articulates that when she was using the centimeter cubes to build the pattern she started with a square in the middle. After she made the square, she noticed that on top of and at the bottom of the square, there is a row of $1 + n$, so there are $2(1 + n)$.

### *Student B*

This student does identify an expression that reaches the correct answer, but there is no correlation to the expression and the way in which the pattern is growing. It is difficult to understand how the student came up with this expression.

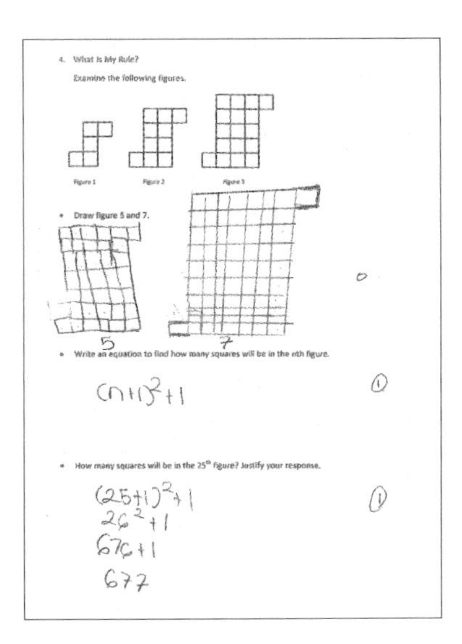


### *Student C*

This student draws an accurate representation, and although she or he does not explain the process used to write the expression, we can see that the student identifies the square in the middle with a rectangle on the top and the bottom plus two additional squares. The expression explains that quite clearly.

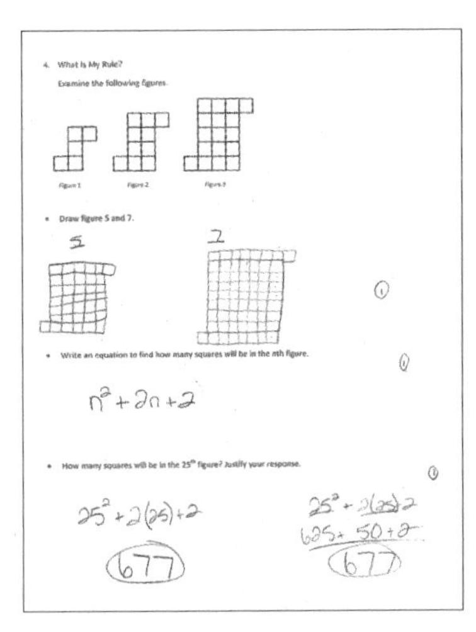


Once the class discussion is complete, this teacher assigns *Growing Figures 2*, page A1-14 in the appendix, for her students' homework.

## Meeting Individual Needs

For many students, it is helpful to begin with a verbal description before writing a mathematical expression describing how the pattern is growing. For instance, "I see a square and two long rectangles." Once the sentence is written, it is easier to write an algebraic statement by substituting numbers and variables immediately under the written words. For instance, if the students describe a 3-by-3 square plus a 2-by-3 rectangle, they can translate that into an algebraic statement as $3 \times 3 + 2 \times 3$. This process is extremely effective for those students who struggle with writing an expression simply by inspection.

## Additional Reading/Resources

Friel, Susan N., and Kimberly A. Markworth. 2009. "A Framework for Analyzing Geometric Pattern Tasks." *Teaching Mathematics in the Middle School* 15 (1): 24–33.

Math Talks. "Pattern Talk 1–4." http://www.mathtalks.net/pt-1-4.html.

Visual Patterns. http:www.visualpatterns.org.

# 4. Taxes and Interest

**DOMAIN: Expressions and Equations**

**STANDARD: 7.EE.2.** Understand that rewriting an expression in different forms in a problem context can shed light on the problem and how the quantities in it are related.



## Potential Challenges and Misconceptions

Many students lack conceptual understanding of how taxes and interest impact the cost of an item. Combine this with their difficulty working with decimals and percents and it is obvious why they have such difficulty with solving tax and interest problems. What students do understand is often lost when they are shown expressions for calculating tax and final cost (or principal, interest, and final cost) without strategies to help them understand what those expressions represent. Students often need help to recognize that when paying taxes, they are actually paying more than 100%. For example, a 6% tax actually means 100% + 6%, or 1.06 times the stated cost. When students develop an understanding of *why* the expression works, they are often able to derive it themselves should they forget or misremember it. By understanding what the *starting value*, *growth factor*, *growth rate*, and *time span* are and how they relate, most students are able to solve most situational problems that include taxes and/or principal, rate, and time.

## In the Classroom

One teacher has determined it is most helpful for his students to begin by describing the situation with a verbal expression and to then use that verbal expression as a guide for their algebraic expressions. He has found this helpful when working with tax and the cost of an item with the tax added and with determining the amount of money in an account once simple interest has been added.

For instance, when thinking about the problem *Jorge bought a shovel for $16. In his state, the sales tax is 6.5%. How much did Jorge pay for the shovel?*, this teacher tells his students to write a verbal expression that describes the situation. Albert volunteers, "I wrote *starting value times a hundred percent plus the percent of the tax.*"

This teacher asks if anyone has a different answer. Tamara volunteers, "I wrote *hundred six and a half percent times sixteen dollars*, because the tax is added to the sixteen dollars. One hundred six and a half percent is 1.065 in decimals, so the expression is 1.065(16)." Again, he asks if anyone did it differently. He encourages all students to share how they are thinking about the situation.

Following the sharing, he tells the students to write the mathematical expression using the values given in the problem and/or to model the situation to check the reasonableness of their answers. After an allotted time, he asks a volunteer to share a diagram followed by an expression. Lily shares her model, a double number line, and her expression, 16(1.00 + 0.065).

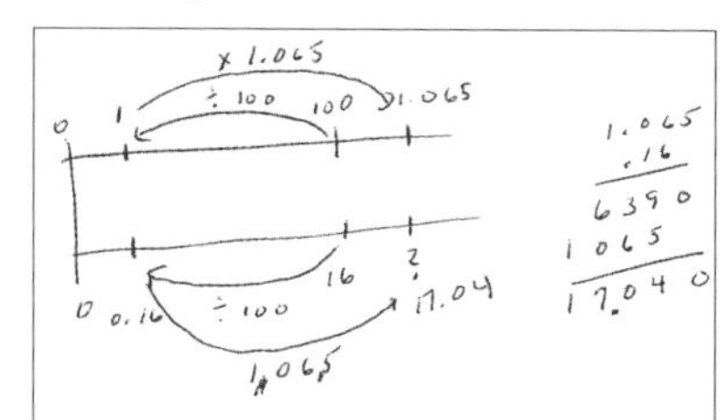


Owen shares his percent table and expression 16(1.065) and explains that he used Tamara's written expression as a guide. He further explains that he used benchmark percents for the taxes. "I saw that with a 10% tax would be a total cost of $17.60 and with a 1% tax, the total cost was $16.16. So I knew the cost with a 6.5% tax would be between the two."

| Value | Tax | Amount of Tax | Cost with Tax |
|---|---|---|---|
| $16 | 100% | $16 | $32 |
| $16 | 10% | $1.60 | $17.60 |
| $16 | 1% | $0.16 | $16.16 |
| $16 | 6% | $0.96 | $16.96 |
| $16 | 0.5% | $0.08 | $16.08 |
| $16 | 6.5% | $1.04 | $17.04 |

Next, this teacher provides graphing calculators and tells students to compare the expressions 16(1.00 + 0.065) and 1.065(16). After they input the two expressions, he tells them to graph those expressions and describe what they see. He notices that some students are confused when only one line appears in the window display. He asks whether anyone can explain what is happening. One student responds, "This is like when we graphed equivalent ratios. All the ratios are on the same ray, so these two expressions have to be equal."

To reinforce understanding the multiple forms in which situations can be modeled and expressions can be written, he pairs his students to play the matching game *My Counterpart* on page A1-15 in the appendix. Note that teachers must cut out the cards before giving them to students. After students play the game by matching verbal models and mathematical expressions, he assigns the reproducible *How Much Will I Pay?* on page A1-17 in the appendix.

All three students would benefit from doing more work with simple percents before moving on to compounding interest rates over time. Students A and B would do well to use 10-by-10 grids and to model how 6% compares to 60%. Both students should also work with percent tables, double number lines, or both to understand what happens to the decimal point when a percent is converted to its decimal equivalent. Student C would benefit from being interviewed to discover how or what he was thinking with solving this problem.

### *Student A*

The student work in this case illustrates an understanding of how to find interest but not how the interest is added to the original cost of the bicycle. This student sets up the equation for calculating interest, correctly identifies the starting value and growth rate, but misunderstands that in year 2 the interest will be applied to $954, not $900. This illustrates either a misconception or misremembering that the principal grows and that the interest in year 2 is calculated on that new amount. In addition, the student operated as though the calculation needed the distributive property. The student multiplied 900 × 2 and added that product to 900 × 0.06. The answer is not reasonable. The bicycle will cost $1,011.24, not $1,854.

1. Sam is planning to buy a racing bicycle that costs $900 for his son. He plans to pay for it over 2 years at a 6% s interest rate compounded annually.

- Write an equation that will help Sam calculate how much he will pay for the bicycle at the end of the two years. Y=900(0.06)+2
- Use your equation to find out how much Sam will end up paying for the bicycle after the two years. $1854

### Student B

Either this student just wrote an expression and used a calculator, or his or her work illustrates a common error with converting rates to factors: failure to convert the percent to a decimal or lack of understanding of how to do so. Instead of dividing by one hundred, or even using the "rule" "drop the percent symbol and move the decimal," this student simply adds a decimal point. The student also neglects to drop the percent symbol, so the new rate is 0.6%, which is a growth factor of 0.006 instead of 0.06. These errors are very common and relate to the students' confusion about percents and their equivalent decimal representations. In addition, this student writes the $900 cost of the bicycle as $9,000. It is difficult to understand where the 1,101 amount comes from without any supporting work. This makes giving feedback to the student difficult.

1. Sam is planning to buy a racing bicycle that costs $900 for his son. He plans to pay for it over 2 years at a 6% s interest rate compounded annually.

- Write an equation that will help Sam calculate how much he will pay for the bicycle at the end of the two years. 9000X.6%
- Use your equation to find out how much Sam will end up paying for the bicycle after the two years.

1011

### Student C

This student work illustrates two major misconceptions. The equation raises the starting value and an incorrect growth rate—1.6 instead of 1.06—to the second power. The student does not check the reasonableness of the multiplication. The final answer is less than the starting value of $900, demonstrating that the student does not check his or her work or has not benefited from writing the verbal model *starting value (growth factor)*$^{time}$.

1. Sam is planning to buy a racing bicycle that costs $900 for his son. He plans to pay for it over 2 years at a 6% s interest rate compounded annually.

- Write an equation that will help Sam calculate how much he will pay for the bicycle at the end of the two years. ($900·1.6)² = Final Price
- Use your equation to find out how much Sam will end up paying for the bicycle after the two years.

(900·1.6)1.6 = F.P.
144·1.6 = F.P.
210.4 = F.P.

When this teacher feels his students can convert decimals to percents and percents to decimals accurately, he assigns the reproducible *Taxes, Interest, and More* on page A1-18 in the appendix.

## Meeting Individual Needs

Invite students who need more scaffolding to make a table using benchmark or friendly percents, as Owen did, or to model the problem using double number lines. To find a tax of 5% on the cost of a toy that costs fifteen dollars, a table and double number lines would look like the following:

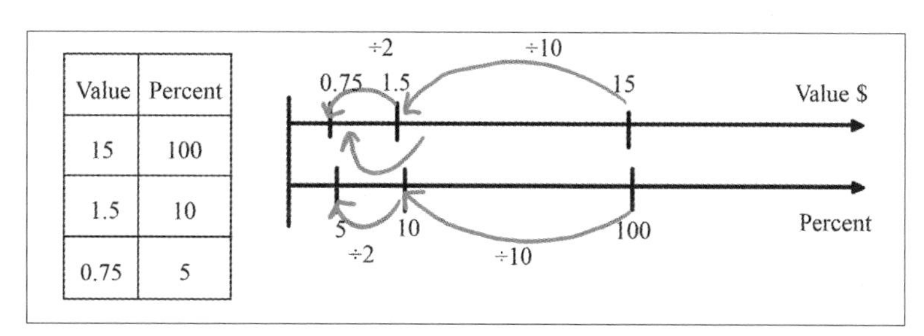

| Value | Percent |
|---|---|
| 15 | 100 |
| 1.5 | 10 |
| 0.75 | 5 |



By reasoning through the fact that the tax must be paid in addition to the cost of the toy, students should realize that they must add the $0.75 to the cost of the toy for a total of $15.75, or they may be guided to see that by dividing 100% by 20 they get 5% and 15 ÷ 20 is 0.75. Be sure to provide students with many practice examples for finding percents. An additional resource is *Taxes, Interest, and More* on page A1-18 in the appendix.

## Additional Reading/Resources

Collins, Anne, and Linda Dacey. 2011. *Zeroing in on Number and Operations, Grades 7–8*. Portland, ME: Stenhouse.

Parker, Melanie. 2004. "Reasoning and Working Proportionally with Percent." *Teaching Mathematics in the Middle School* 9 (6): 326–333.

# 5. Venn Diagrams and Factor Lattices

**DOMAIN:** **Expressions and Equations**

**STANDARD:** **8.EE.7b.** Solve linear equations with rational number coefficients, including equations whose solutions require expanding expressions using the distributive property and collecting like terms.



## Potential Challenges and Misconceptions

Many students are challenged when it comes to determining the common denominators or least common multiples (LCMs) for rational expressions with variables. This is partially due to their discomfort with fractions in general and because many students have been told to cross-multiply. Students need to develop conceptual understanding of the fact that the least common denominator is the smallest shared multiple of both values in the denominators. This can be shown on the Cartesian coordinate plane with numerical values or with Venn diagrams and factor lattices with either numeric or variable factors.

On the Cartesian coordinate plane, the LCM is the first time both rational expressions share the same $x$-value on the grid. For Venn diagrams, the LCM is the product of all factors remaining in each region of the diagram *after* shared values have been moved to the intersecting regions. For factor lattices, the LCM is the first vertex at which the two operators intersect. Students who have experience with these models using numeric values demonstrate less difficulty when working with expressions with variables.

## In the Classroom

Calculating the LCM or greatest common factor (GCF) is challenging when working with numbers but much more difficult when working with variables and exponents. Students who have multiple ways of thinking about these concepts tend to succeed, whereas those who rely on algorithms struggle. Two effective strategies include Venn diagrams and factor lattices. Students who have experience with both tend to demonstrate a deeper understanding of factoring, prime factorization, LCM, and GCF.

One teacher first introduced the Venn diagram and factor lattice strategies when working on these concepts with numbers in the middle grades. When she sees how her algebra students struggle with the same concepts when variables are introduced, she decided to use the same two strategies.

### Venn Diagrams

This teacher begins by projecting a Venn diagram (see *Venn Diagrams* on page A1-19 in the appendix) and the expressions $x^4y^3$ and $x^5y^6$ on the board. She challenges the class to write each term in expanded form on sticky notes, with one factor per note. She asks students to represent the terms by placing the sticky notes on the Venn diagram and identifying the GCF and the LCM. After an allotted time, she invites student volunteers to share their diagrams and show how they used sticky notes to arrive at their results.

Carter volunteers to share his diagram for finding the GCF. As he explains what he has done, he models his work using the sticky notes. He places all the factors for each term in the outer parts of the Venn diagram. Then he carefully moves pairs of factors common to each term (or shared by each term) into the overlapping section, placing one sticky note of the pair on top of the other. He explains, "Multiply all the stickies in the overlapping part (or intersection) to get the greatest common factor."

Students discuss why you don't multiply the factors hidden under other sticky notes. Myles sums up the discussion by saying, "If you want the greatest common factor, you multiply only the factors in the shared or common section of the Venn diagram. Because the factors are shared, you only have to multiply them once. You just ignore the factors in the other parts of the circles, because they are not shared."

To find the LCM Anne explains, "We know a multiple means you get a product. So, to get a multiple of the term on the left, you multiply all the factors in the whole left circle by any factor or factors from the right circle. You do the same thing to get a multiple of the term on the right—multiply the factors in the whole right circle by any factor or factors in the left circle. But to get a multiple of both terms, you have to multiply the factors in all three sections—left, middle, and right." Students discuss Anne's explanation, and when most students seem to understand, the teacher asks them to model one more. She posts $6xy^3$ and $4x^2y$. After a short time, Aliana shares her diagram.

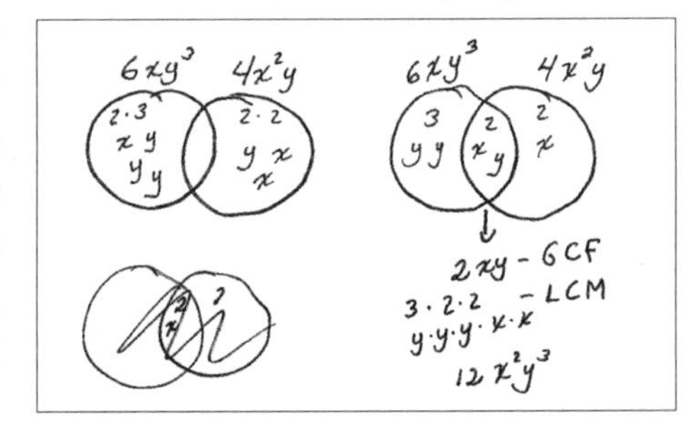


Again, this teacher asks whether anyone got anything different or did it differently. When no student has anything to add, this teacher assigns the *Venn Diagram Representations* reproducible from page A1-20 in the appendix.

### Factor Lattices

A factor lattice is another visual representation that helps students understand GCF and LCM. Because her students have used factor lattices previously, the teacher moves directly to the pictorial representation. She asks her students to draw a factor lattice for the term $x^2y^3$. Her students know to define each factor and to determine and define in which direction each factor will assume. After an allotted period of time, she invites Donovan to share his diagram.

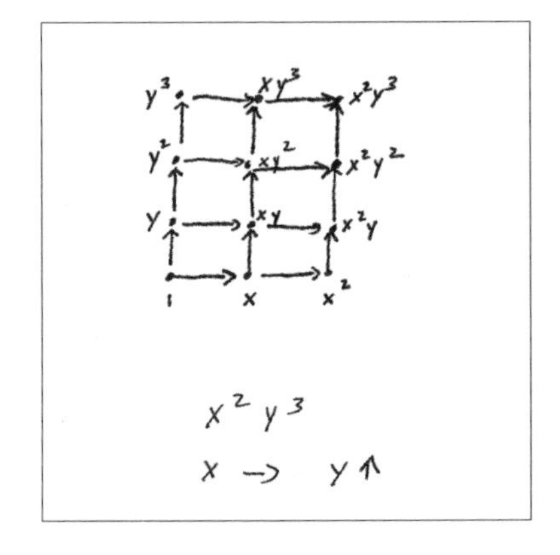


Once Donovan shares his representation, this teacher asks all students to check their factor lattices to be sure they are accurate. She then explains that she will show students how to use the factor lattice to find the LCM of $x^2$ and $y^3$. She challenges them to figure out how they can use this method for any two terms. The teacher asks Donovan to put one finger on $x^2$ and the second finger on $y^3$ and to trace the pathways to the right and up until they intersect. She explains that the point of intersection, $x^2y^3$, is the LCM of $x^2$ and $y^3$. She asks the other students to do the same and to begin a table to record the first term, the second term, and the LCM.

This teacher dictates a few more pairs of factors and encourages different students to go to the document camera and show how they might find the LCM. For example, she tells them to put their fingers on $y^2$ and $x^2y$ and trace their fingers right and up to find the LCM ($x^2y^2$) and to put their fingers on $x$ and $y^3$ to find the LCM, $xy^3$. For each example, she has all students record the terms and the LCM in their tables.

| Term | Term | LCM |
|---|---|---|
| $x^2$ | $y^3$ | $x^2y^3$ |
| $y^2$ | $x^2y$ | $x^2y^2$ |
| $x$ | $y^3$ | $xy^3$ |

After determining the students are gaining proficiency, this teacher tells her students to make a conjecture about how they might find the LCM between or among any given variables. Although some of the students suggest that the LCM is the product of the variables with the smallest exponents, the majority of students contradict that suggestion and state, "It is the product of the variables with the largest exponents. This teacher then instructs her students to use the examples in their tables to determine which of the two conjectures is supported by the data. It would have been easy for her to tell her students the rule, but she wanted her students to use their tables to justify their responses (SMPs 3, 7).

After discussing the LCM, this teacher moves on to using the factor lattices to determine the GCF, or greatest common divisor. This requires students to place their fingers on two terms and to move to the left and down to the point of intersection. For example, if trying to find the GCF for $y^2$ and $x^2y$, students trace back to $y$. As previously, she dictates examples and has students record their results in a table.

| Term | Term | GCF |
|---|---|---|
| $y^2$ | $x^2y$ | $y$ |
| $x^2y^2$ | $y^3$ | $y^2$ |
| $xy^2$ | $x^2y$ | $xy$ |
| $x$ | $y^3$ | 1 |

She asks her students to make a conjecture about how to find the GCF given two or more variables. This time, all her students agree that the GCF is the variable with the smallest exponent. After working with many examples, this teacher assigns the reproducible *Factor Lattices 1* on page A1-21 in the appendix.

SECTION 1: EXPRESSIONS

## Meeting Individual Needs

When modeling the Venn diagrams, it is extremely beneficial to put each factor on a sticky note. When there is a common factor, students can place one sticky note on top of another in the region of intersection to help them understand that no values were canceled out. The stacking illustrates a sharing of that specific factor.

Many students better understand the factor lattices after they have built them using play dough or modeling clay (or gumdrops if candy is allowed in class) as the vertices and toothpicks as the operators. As the students build their models, they are able to trace the pathways from one factor to another and to physically see where the pathways intersect. When moves are made up and to the right, the point of intersection is the LCM; when the moves are made down and to the left, the point of intersection is the GCF. Students should also be encouraged to draw the factor lattice as a reference when working with rational expressions.

## Additional Reading/Resources


Bezuszkas, Stanley J., and Margaret Kenney. 2001. *Number Treasury 2: A Collection of Facts & Conjectures, Problems & Investigations, About More Than One Hundred Kinds of Numbers* (Grades 7 & Up). Palo Alto, CA: Dale Seymour Publications.

# 6. Area Expressions

**DOMAIN: Expressions and Equations**

**STANDARDS:** **6.EE.7.** Solve real-world and mathematical problems by writing and solving equations of the form $x + p = q$ and $px = q$ for cases in which $p$, $q$ and $x$ are all nonnegative rational numbers.

**7.EE.2.** Understand that rewriting an expression in different forms in a problem context can shed light on the problem and how the quantities in it are related.

## Potential Challenges and Misconceptions

Area and perimeter present challenges for students from their first introduction in the elementary years. Many students do not understand that a rectangle has two dimensions: length and width. In the middle school years, writing expressions for the area of a figure whose dimensions include a variable that relates width to length causes great difficulty for many students. This difficulty arises due to having only previous experiences that relate area to the algorithm $A = l \times w$. Students are usually successful when the dimensions are given and all that is required is to "plug and chug" numbers. However, oftentimes one dimension and the area or perimeter are given and students are required to calculate the missing value—for instance, *the length is 9 inches and the area is 36 square inches, find the perimeter*. These multiple-step situations require a conceptual understanding of both area and perimeter that, unfortunately, not all students have. It is vital that students have multiple experiences modeling figures and writing expressions and equations. When the dimensions include variables that relate one dimension to the other, the experiences provide an introduction to multiplying binomials.

## In the Classroom

One teacher begins her lesson on area and perimeter by projecting a square on which the width is labeled. She asks her students how they might find the perimeter. Giana states, "We could use a piece of string and wrap it around the square, then measure how long the string is." Brett interjects and states, "That only works if we have numbers. We can't just measure with string when a variable is used as the length of a side. We need to think differently." The students work in pairs to find the perimeter.

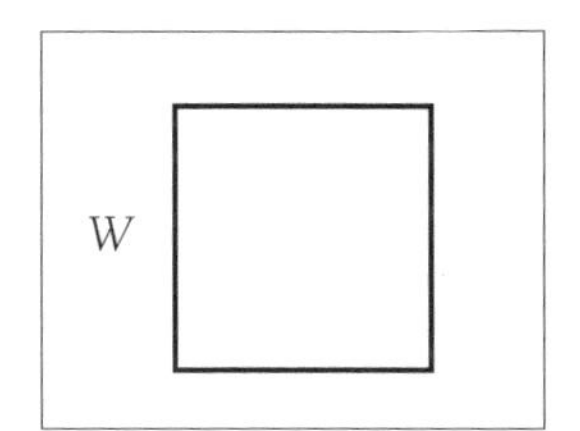


After an allotted time, students share their work. Carter shows his work and explains, "I just added up all the sides." Carter demonstrates conceptual understanding, but he has only listed three side values rather than four.

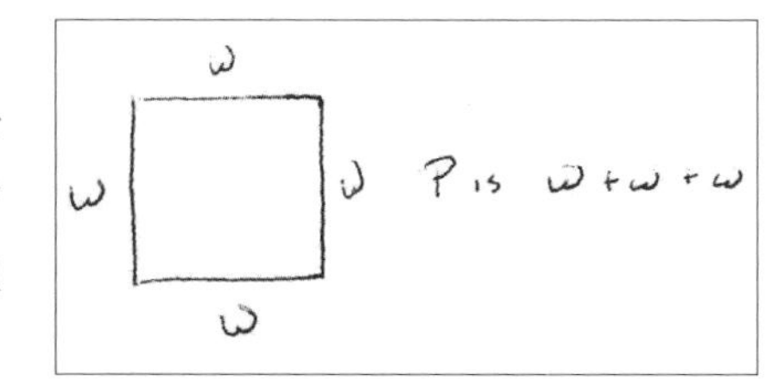

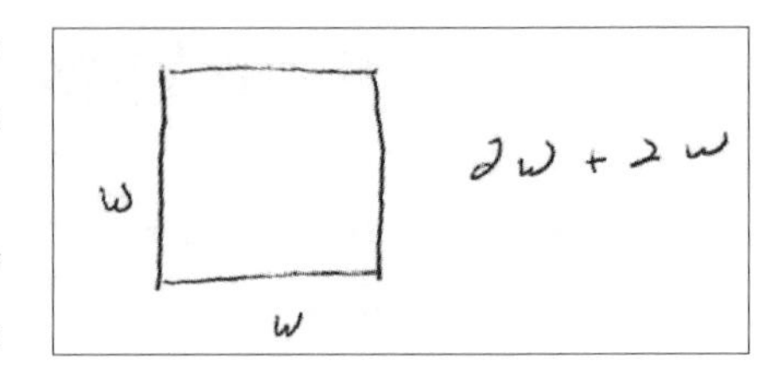


Jacob shares his work and explains, "There are two opposite sides, so I just multiplied each side by 2. Then I could add them together and get 4*w*."

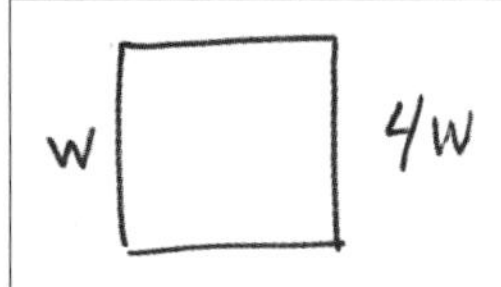


Quinn explains, "I did mine like Jacob, but I saw that if one side of a square is *w* then all four sides have to be the same. So I got 4*w*."

Next, this teacher projects a rectangle and tells her students that the length is 3 more units than the width. She challenges her students to determine the area. She instructs her students to work in pairs on this problem. After an allotted time, she invites volunteers to share their models.

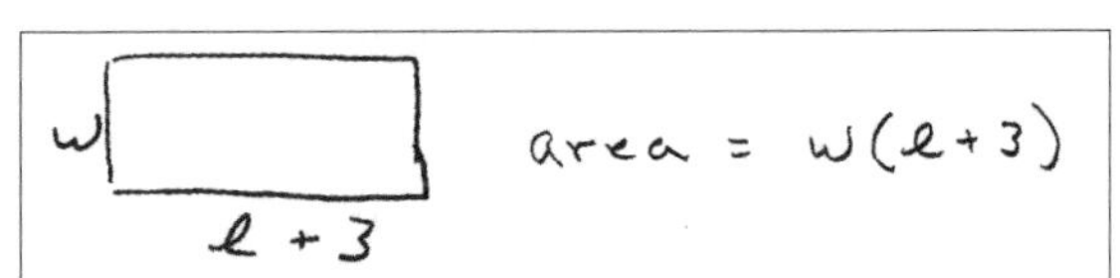


Marina shares her work. Notice that she uses two unique variables rather than illustrating the relationship between the length and the given width.

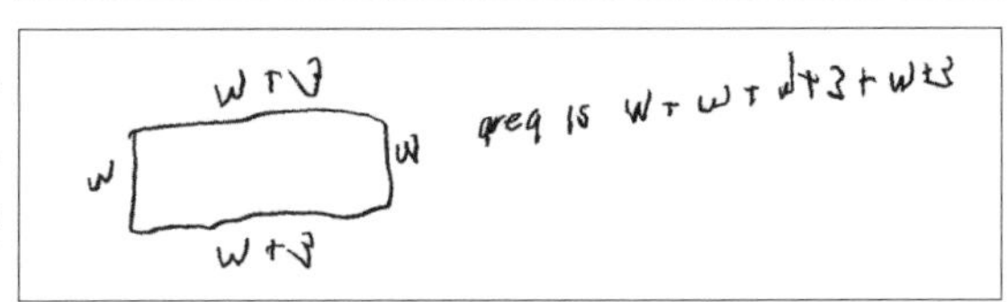


Next, Henry volunteers to display his work. Henry makes the most common error students make when working with area: he confuses area with perimeter. He does illustrate, however, an understanding of how to label the length in relation to the width.

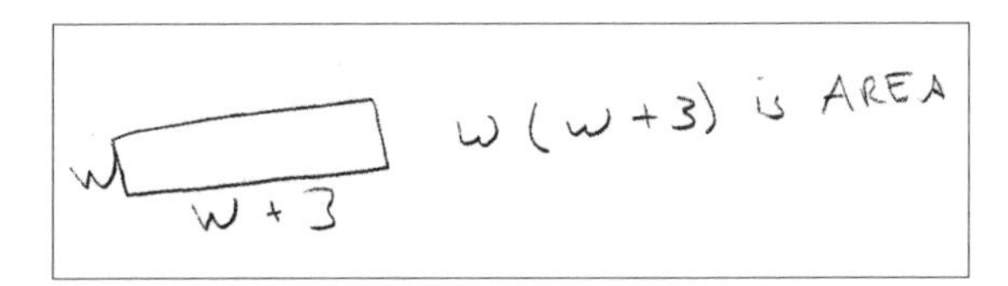


Ricardo correctly identifies the length in relation to the width and uses the appropriate algorithm for finding the area. He asks if any of the students have any questions for him. This teacher confirms that Ricardo's expression is accurate and challenges them to write an expression equivalent to $w(w + 3)$. She reminds them that this expression is in factored form with width ($w$) as one factor and the quantity ($w + 3$) as the second factor.

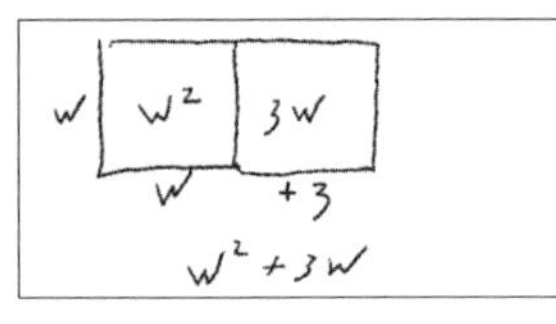


Lucy shares her equivalent expression $w^2 + 3w$. She explains that she had to use the distributive property. "If you look at the drawing, you can see the first part is the square with dimensions *w* by *w* and the second part is 3 units long by *w* units wide for 3*w*. When you put them together you get $w^2 + 3w$."

This teacher then asks the students to write two equivalent expressions for the perimeter of the same rectangle. When the students are finished, she tells them to compare their expressions with a shoulder partner. Following that brief discussion, she asks student volunteers to share their expressions.

### *Student A*

The evidence in this student work demonstrates an understanding that the sum of the measures of the four sides comprises the perimeter. The work also illustrates an understanding of how to express one dimension in terms of another and how to combine like terms. The expression is accurate and represents an expression for finding the perimeter.

### Student B

The evidence in this student work demonstrates the fact that this student may remember the formula $p = 2(l + w)$ or possibly an understanding of how to use the distributive property when doubling the sum of the width and the length of the rectangle. It also shows an understanding of how to combine like terms. It would be interesting to ask this student to write a third expression to determine whether she or he would apply the distributive property to the second expression for a product of $4w + 6$.

w

w + 3

2(w + w + 3)

2(2w + 3)

### Student C

The evidence in this student work illustrates an understanding of the formula for finding perimeter as $2w + 2l$. The second expression also shows an understanding of the distributive property. Again, it would be interesting to ask this student how he or she might write a third equivalent expression by combining like terms.

w

w + 3

2w + 2(w + 3)

2w + 2w + 6

This teacher asks the class to think about the difference between the area and perimeter of figure and records their responses on a conjecture board. The conjectures included: "Perimeter is the distance around the edges," "Perimeter is the outside and area is the inside space," "Area is space," and "Perimeter is a line." She tells the class that another student included, "Area is two-dimensional and you multiply the two dimensions to find the area of a rectangle, but perimeter is one-dimensional and you add the dimensions."

This teacher uses the conjecture board as a way of determining what her students know about a particular concept or topic. Because a conjecture is an idea that needs to be proven, the conjecture board is a safe place where students can share their thinking whether correct or not. It also embraces SMP 3: Construct viable arguments and critique the reasoning of others.

After recording the conjectures, this teacher projects another rectangle and challenges her students to find both the area and the perimeter and to write those expressions in at least two ways.

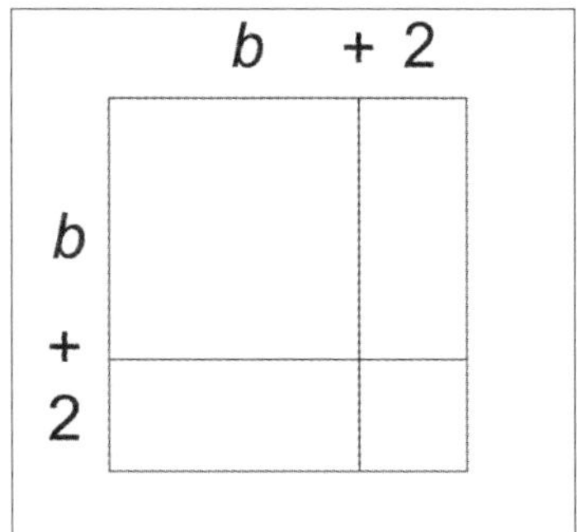


This teacher provides easel-sized paper to groups of three with instructions to record their work. She gives her students Algeblocks to model the expression. Once the students have had time to model and write at least three expressions, she tells them to post their work around the room for a facilitated gallery walk. The gallery walk is another way in which this teacher incorporates SMP 3: Construct viable arguments and critique the reasoning of others. To guide the students, she tells them to look at similarities and differences in the expressions. She asks students looking at the work to make notes on sticky notes and post them if they have questions, need clarification, or just disagree with the expressions. After completing the gallery walk, this teacher assigns the *Modeling Algebraic Expressions* from page A1-22 in the appendix.

## Meeting Individual Needs

One of the most effective ways to meet the needs of students who struggle with these concepts is to use Algeblocks together with the factor grid and four-quadrant mat. These manipulatives allow students to place the factors in a factor grid and build the area in one of four quadrants. These concrete manipulatives are effective in assisting students in building both factors and products for expressions and are especially powerful in modeling quadratics.

## Additional Reading/Resources

Collins, Anne, and Linda Dacey. 2011. *Zeroing in on Number and Operations, Grades 7–8*. Portland, ME: Stenhouse.

ETA hand2mind. "Algeblocks." http://www.hand2mind.com/brands/algeblocks.

Kaplinsky, Robert. 2015. "Why Depth of Knowledge Is Critical to Implement." February 23. http://robertkaplinsky.com/why-depth-of-knowledge-is-critical-to-implement/.

# 7. Multiplying Polynomials

**DOMAIN: Algebra**

**STANDARD: A-APR 1.** Understand that polynomials form a system analogous to the integers, namely, they are closed under the operations of addition, subtraction, and multiplication; add, subtract, and multiply polynomials.



## Potential Challenges and Misconceptions

Too often, in an attempt to help, adults tell students what to do and how to do it, or offer mnemonic devices, without paying attention to the conceptual mathematics behind all those instructions. For example, ask most adults how to multiply a binomial by a binomial and the majority will say, "FOIL it." The ubiquitous acronym FOIL can be a useful mnemonic device when multiplying two binomial expressions, but the device is useless when multiplying expressions that are not binomials or when there are more than two binomial expressions to multiply. The distributive property, the more general mathematical law that FOIL exploits, is relatively straightforward to use without mnemonic tricks. It is important to emphasize the distributive property rather than using the mnemonic alone if we are to help students develop understanding and prevent calculation errors.

## In the Classroom

One teacher begins his lesson on the importance of using the distributive property when multiplying binomials or polynomials by posing a range question. Recall that a range question is designed to let the teacher understand what students know about a particular concept, what they misremember, or what misconceptions they bring to the concept. He asks, "Tell me what you know about the expression $(a + b)^2$ and give me an equivalent expression for it."

Darlene excitedly reports, "That is really easy! You get $a^2 + b^2$."

Hannah disagrees and states, "I would write the equivalent expression as *the quantity of* a *plus* b *times the quantity of* a *plus* b, *or* (a + b)(a + b)."

Jaxon adds, "My brother told me that you just FOIL the numbers or variables after you rewrite the expression like Hannah did. So my equivalent expression will be $a^2 + 2ab + b^2$." This teacher records each response on the board. He asks if anyone has anything to add before he distributes Algeblocks and quadrant mats to each student. He tells students to model the original expression to determine which responses are correct. He also tells students to sketch the Algeblock model in their notebooks.

After an allotted time, this teacher calls on student volunteers to share their mats. Lionel shows his first. He explains that he used the yellow rod for the $a$ value and the orange rod for the $b$ value. He built the first factor $(a + b)$ horizontally, and he built the second factor, also $(a + b)$, vertically.

Then Lionel explains, "When you are multiplying any two numbers, you can show the product as a rectangle. And I got a rectangle. My answer is $a^2 + 2ab + b^2$." Before sitting down, he exclaims, "Isn't it cool that when you multiply $a$ by $b$ you get $ab$, but look at the blocks—yellow times orange gives tangerine! The $a$ and $b$ blend together. That is just so cool!"

This teacher looks at Darlene to see whether she understands Lionel's reasoning. She points to her own quadrant mat, which shows the same—correct—model, and smiles. He then distributes the reproducible *Expanding Products 1*, page A1-23 of the appendix, and asks students to work on the first question, $(a + 2)(a + 3)$. When most students are ready, he invites one student to share a solution with Algeblocks and another student to share her sketch with the document camera:

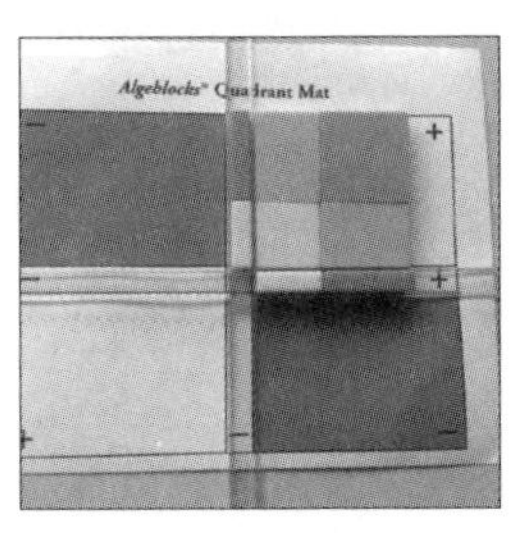

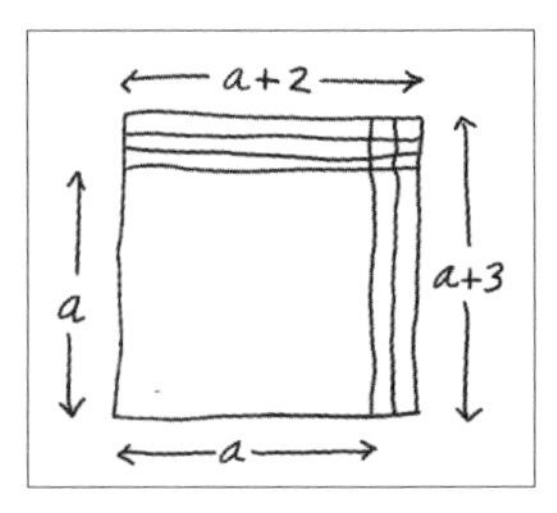


Tamara explains her sketch, "I made an outline of an $(a + 2)$-by-$(a + 3)$ rectangle; then filled in an $a$-by-$a$ square, three $a$-by-1 rectangles on top, two 1-by-$a$ rectangles on the side, and 6 unit squares in the upper right corner. Adding up the areas, I got $a^2 + 3a + 2a + 6$, which equals $a^2 + 5a + 6$."

Jamaal says that he thought about drawing a picture and knew it would look like Tamara's. "I knew I'd get an $a^2$, two $a$'s, three more $a$'s, and a 6, which adds up to $a^2 + 5a + 6$, but I wanted to build it to be sure." He continued to work with the Algeblocks.

Jaxon asks, "Can't you just FOIL it? That's what I did."

$(a+3)(a+2)$
$a(a+2) + 3(a+2)$
$a^2 + 2a + 3a + 6$
$a^2 + 5a + 6$

Heather says, "I can never remember what FOIL means, so I distribute. It's easier." She writes on the board and explains: "See, I distributed the $a$ over the $a + 2$, and then distributed the 3 over the $a + 2$." She points to the third line. "First, I got $a^2 + 2a$. Then I got . . . no. I made a mistake. I really got $3a + 6$, not 2." She corrects her error. "I simplified to $a^2 + 5a + 6$. You need to be careful to distribute each factor in the first parenthesis over both terms in the second parenthesis. I remember hearing the distributive property is really the 'distributive property of multiplication over addition.'"

"That's exactly the same as I got with Algeblocks and what Tamara's picture shows: $a^2 + 5a + 6$!" Nikolas remarks. The teacher explains that the blocks and sketch are modeling the product, so it's not a surprise that the results look the same. He then asks students to pair up and talk about what they understood and what difficulties they had with the calculations they had done. He finds that his students do a good job of helping each other understand new concepts and methods, so he spends a lot of time in class having them work in pairs or triads modeling various computations.

This teacher asks whether anyone has any questions. When no one does, he moves on to having his students multiply a trinomial by a binomial: $(a^2 + 2a + 1)(a + 1)$. As they work, he walks around the room observing how they are approaching this computation.

He observes Oliver rewriting the trinomial as two binomials $(a + 1)(a + 1)$ and building it with Algeblocks. When asked what he is thinking, Oliver responds, "Well, I know that $(a^2 + 2a + 1)$ is the product of $(a + 1)(a + 1)$, but it is also a factor, so I have to multiply it by the other $(a + 1)$. I think that I am going to get a cube, because all the factors are the same and there are three factors." This teacher notices Oliver is using several mathematical practices, including SMPs 1 and 3.

He next asks Oliver how he might model the trinomial as a factor. Oliver looks puzzled but tries two different strategies. Oliver's models are shown below. Picture A represents $(a + 1)^2$, and Pictures B and C show his first and second attempt at modeling $(a + 1)^3$.

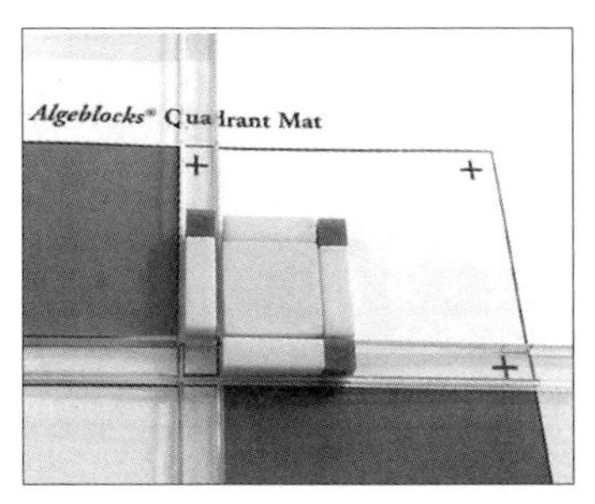

Picture A

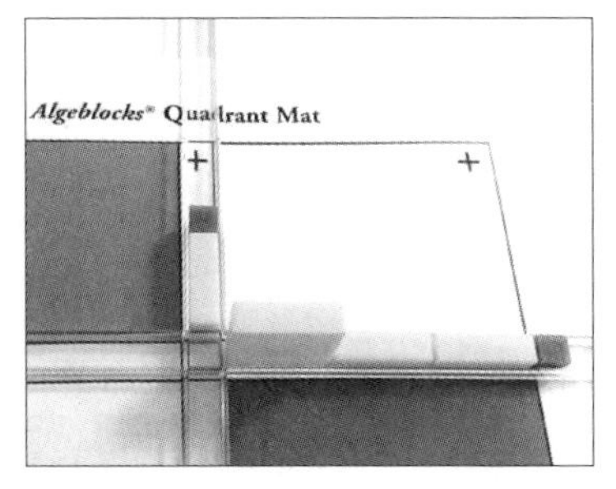

Picture B

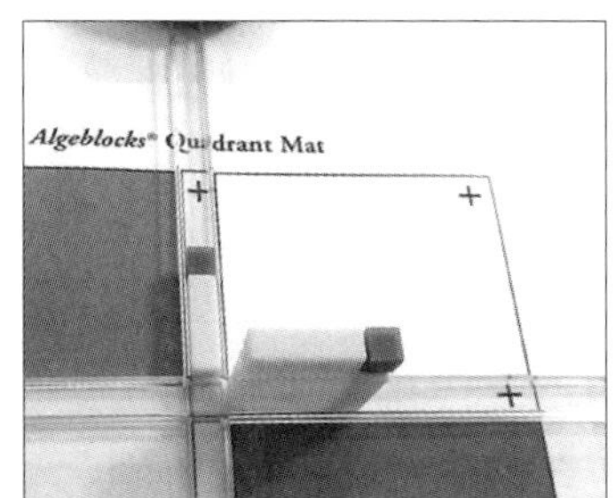

Picture C

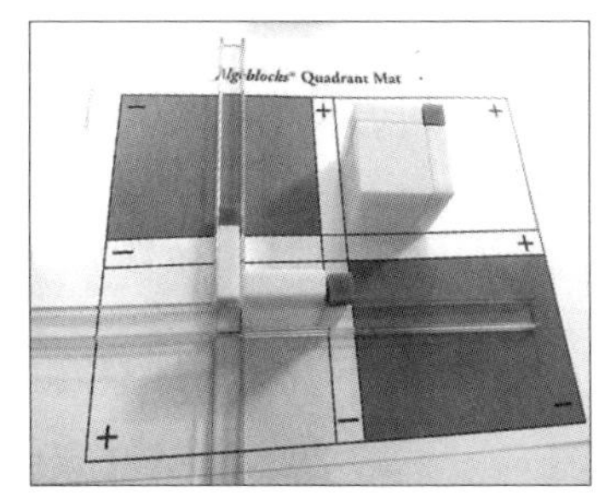
Picture D

Picture D shows both the factors and the cube for $(a + 1)^3$.

Oliver explains, "First I tried to line up the tiles from my area model, but that didn't work. So I balanced the array in the factor track and worked building the cube from there. This worked, and I ended up with an array as my factor."

Joey was sitting by himself trying to figure out how he might FOIL the trinomial with a binomial. This teacher sat down beside him and watched him struggle. After a while, he asked Joey what he was thinking. Joey responded, "My brother told me that FOIL always works, but I can't figure out why I can't make it work here." This teacher asked Joey to write the trinomial and underneath it to write the binomial. He wrote $a^2 + 2a + 1$. His teacher told him to do the multiplication just as he would with numbers.

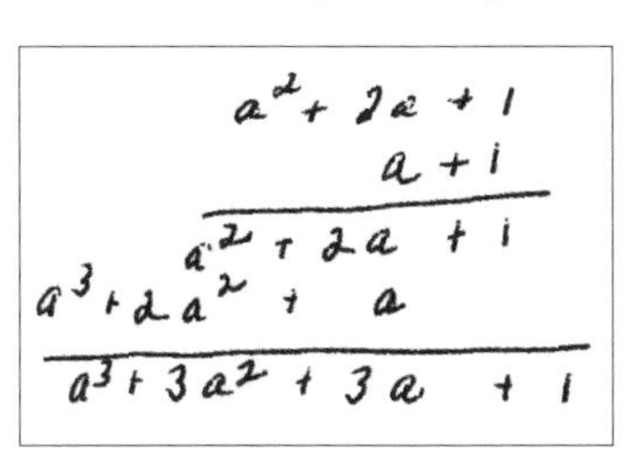


As students start to work on the next problem, *Multiply* $(a + 3)(a^2 + 5a + 6)$, there is discussion about the fact that the Algeblocks wouldn't work. Although that is actually not the case (as cubic polynomials *can* be represented with the blocks), this teacher doesn't intercede because his students found ways to work on the problem. This teacher also notes that many other students still prefer changing all variables to *x* rather than working with the variable *a*.

Tamara says, "I'm going to try to use Heather's method" and writes in her notebook:

$$( (a+3)(a^2+5a+6)$$
$$( \quad a(a^2+5a+6) + 3(a^2+5a+6)$$
$$a^3 + 5a^2 + 6a + 3a^2 + 15a + 18$$

"Oh, now I see why it is the distributive property of multiplication over addition, like Heather said." Jamaal works with her and adds, "Now we can combine terms with the same power of *a*. First, $5a^2$ and $3a^2$ add up to $8a^2$. Then, $6a$ and $15a$ add up to $21a$. That means the answer is $a^3 + 8a^2 + 21a + 18$." His teacher is impressed that the students are not adding equal signs to the expressions or using strings of equal signs, a misconception that he has been addressing since the first day of school.

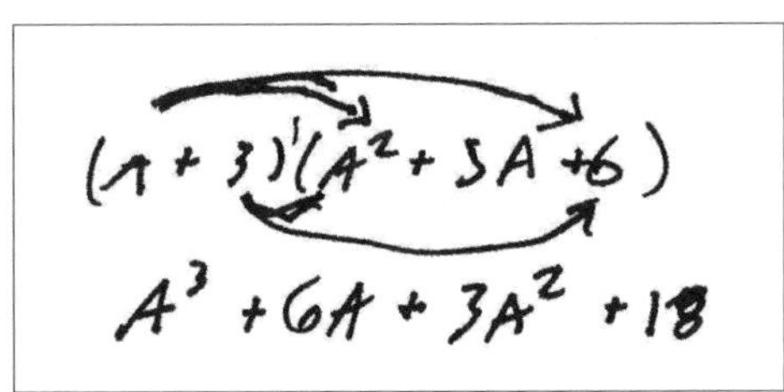


Joey holds up his paper and says, "That's not what I got. I FOILed it and got this, but that's not what Tamara and Jamaal got!"

Heather moves over next to Joey. She works with him to see that FOIL doesn't work, because the second factor ($a^2 + 5a + 6$) has three terms instead of two, and FOIL is set up to work with two expressions with *only* two terms each. She tells him, "That's another reason not to use FOIL—it only works sometimes! Just think of multiplying each term of each sum together and adding them up. That's the distributive property—it always works when you have to multiply by sums."

$$(a+3)(a^2+5a+6)$$
$$(a+3)(a+3)(a+2)$$
$$(a+3)^2(a+2)$$

Nikolas excitedly adds, "I just noticed that $a^2 + 5a + 6$ is the same as $(a + 3) \times (a + 2)$. We just computed $(a + 3) \times (a^2 + 5a + 6)$, so that's the same as $(a + 3) \times (a + 3) \times (a + 2)$, which is $(a + 3)^2 \times (a + 2)$!" The teacher records Nikolas's ideas on the board and then challenges the class to expand $(a + 3)^2 (a + 2)^2$.

Nikolas quickly starts working and says, "We just have to multiply $a^3 + 8a^2 + 21a + 18$ by $(a + 2)$. You can distribute the same way, even though the binomial is written last, after that other long factor." Working in pairs, the class labors on the calculation:

$$(a^3+8a^2+21a+18)(a+2)$$
$$(a^3+8a^2+21a+18)a + (a^3+8a^2+21a+18)2$$

Not every pair gets the correct answer the first time, mostly because of computation errors. But with patience and perseverance, the class arrives at the same correct solution. The students discover for themselves that their computation errors are affecting their solutions. After completing their discussion, this teacher assigns the reproducibles *Expanding Products 2*, page A1-24 in the appendix, and *Don't Get FOILed Again!*, page A1-25 in the appendix.

## Meeting Individual Needs

Many students are able to model multiplication of binomials by binomials using Algeblocks or by sketching the model, which is sometimes referred to as "drawing window panes." Some students do extremely well multiplying polynomials by polynomials by using lattice multiplication. This is particularly true for students who used the Everyday Math curriculum in the elementary grades. Everyday Math Grade 5, and other elementary curricula, offers multiple strategies for multiplication, including partial products and lattice multiplication. Adapting lattice multiplication with integers to working with variables and integers is a seamless transition, and often students only need to practice doing the lattice multiplication examples two or three times to refresh their memories.

This is the lattice for $(a + 3)(a^2 + 5a + 6)$:

| | $a$ | $+3$ | $\times$ |
|---|---|---|---|
| | $a^3$ | $3a^2$ | $a^2$ |
| $a^3$ | $5a^2$ | $15a$ | $+5a$ |
| $8a^2$ | $6a$ | $18$ | $+6$ |
| $21a$ | $18$ | | |

The factors are written in the gray cells, the partial products are in the center cells, and the gray terms make up the final product. Notice that after all the partial products are written, the addition occurs on the diagonals, which makes combining like terms easy to see. The product is $a^3 + 8a^2 + 21a + 18$.

## Additional Reading/Resources

Benson, Steve, Susan Addington, Nina Arshavsky, Al Cuoco, E. Paul Goldenberg, and Eric Karnowski. 2004. *Ways to Think About Mathematics: Activities and Investigations for Grade 6–12 Teachers*. Thousand Oaks, CA: Corwin.

# 8. Powerful Integers

**DOMAIN: Expressions and Equations**

**STANDARD: 8.EE.1.** Know and apply the properties of integer exponents to generate equivalent numerical expressions.

## Potential Challenges and Misconceptions

Operating with exponents provides a challenge for many students. This is often because students memorize a set of rules without understanding where those rules come from or how they look when modeled. Another challenge arises when students are asked to compute $-4^2$ versus $(-4)^2$. The first example, $-4^2$, requires students to raise four to the second power and then to multiply that product by negative one. This is true because of the order of operations, which specifies working with exponents before multiplying. The second example, $(-4)^2$, requires the integer negative four to be raised to the second power. To help students meet the challenges of exponents and avoid misconceptions, it is important to provide all students with time to model operations with powers of two and three. This is especially important to help students "see" the difference between, say $2^3$, which is three-dimensional, and $2 \times 3$, which represents three groups of two.

## In the Classroom

Helping students understand exponents, how to operate with them, and what the conventions or algorithms for exponents are can be complicated. This teacher believes that the more he requires his students to discover and write rules for multiplying, dividing, and raising powers to powers, the better they understand and are able to remember, derive, and apply those procedures. In addition, he knows that, in discovering the rules from a set of examples, students engage in SMP 7: Look for and make use of structure.

To begin the discussion about exponents and how to operate with exponents, this teacher pairs his students and assigns the *Operating with Exponents* reproducible from page A1-26 in the appendix. When students share their results and explain their reasoning, this teacher interjects the terms *base* and *exponent* as appropriate, modeling precise and clear communication. He notices that his students soon begin using the terms correctly as well. This teacher believes it is best to use the terms *base* and *exponent*, or *power*, repeatedly before expecting his students to embrace them. Once the students demonstrate confidence in what the terms mean, this teacher adds the words to his word wall. The following student work illustrates how students were thinking about computing with exponents.

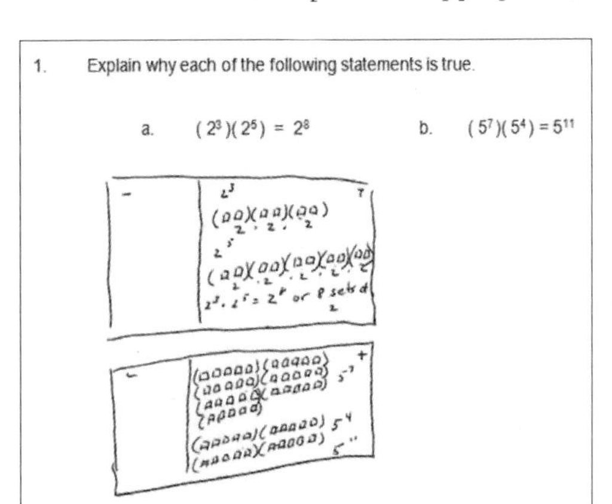


### *Student A*

This student offers this explanation for multiplying expressions with the same base and different exponents. "I drew an algebra mat and modeled the expression and its answer. When you raise two to the third power and multiply it by two to the fifth power, you can see how there are eight sets of two altogether. But then I saw that was wrong. Eight sets of two would be 8 × 2, or 16. I was showing multiplication, not raising to a power, so I wrote out 2 × 2 × 2 × 2 × 2 × 2 × 2 × 2 and got a way bigger answer. I got 256. So really it is $2^8$. For b, I just wrote out the multiplication."

In response to the second item, this student offers, "When I built the expressions from number 1, I noticed that the base stayed at two and I just added the exponents. So for number 2, I got $a^{m+n}$. I kept the base and added the exponents. This matches the model I made for $2^3 \cdot 2^5 = 2^{3+5}$, which is $2^8$."

2. Complete the following equation to show how you can find the exponent of the product when you multiply two powers with the same base. Justify your reasoning.

$(a^m)(a^n) =$ ______

$2 \cdot 2 \cdot 2 \cdot 2 \cdot 2 \cdot 2 \cdot 2 \cdot 2$

$2^8$

$A^m \cdot A^n = A^{m+n}$

### Student B

This student explains that he listed the number of integers as many times as the exponent stated. Then he says that he counted all the integer factors to get the answer. "The 2s stay the same and now there are eight of them, so I think you keep the $a$ and add the $m$ and $n$. My answer will be $a^{m+n}$."

$2 \cdot 2 \cdot 2 \cdot 2 \cdot 2 \cdot 2 \cdot 2 \cdot 2$

$2^8$

$A^m \cdot A^n = A^{m+n}$

### Student C

"I did the same thing as the others, but I think the rule is 'If the bases are the same, keep the base and add the exponents.'"

When reporting answers to number 3 in the *Operating with Exponents* reproducible, many of the students volunteer to share their solutions. This teacher is pleased that the students who shared their solutions all report the same rule: "If the bases are different but the exponents are the same, multiply the bases and keep the exponents."

3. Explain why each of the following statements is true.

a. $(2^3)(3^3) = 6^3$ b. $(5^7)(6^7) = 30^7$

### Student D

"First I listed out the 2 three times and then the 3 three times. I saw that, if I changed the order, then I could write 2 · 3 three times, which is $6^3$."

$2 \cdot 2 \cdot 2 \cdot 3 \cdot 3 \cdot 3$

$2 \cdot 3 \cdot 2 \cdot 3 \cdot 2 \cdot 3$

$6 \cdot 6 \cdot 6$

$6^3$

Before moving on, this teacher challenges his students to model how $3^2$ differs from $3^3$. He provides groups of students with unit cubes and the Algeblock factor track and grid and tells them they have five minutes to complete the task. This teacher believes doing this hands-on model will help them understand the meaning of the exponents while at the same time engaging with SMP 7: Look for and make use of structure.

Savannah shares her work and explains, "I know that the area of a square is side times side, so I got 3 × 3 = 9… oops, I wrote 3 instead of 9." She corrects her work. "And I know that there are two dimensions. When I multiply by another 3, I know it will make a three-dimensional figure, so the length is 3, the width is 3, and there are 3 layers high. So 3 × 3 × 3 = 27. With my Algeblocks I can build a cube for 3 × 3 × 3 but just a square for 3 × 3."

D'Jon adds, "If we multiplied by another 3, it would give us a hypercube, but I don't know how to draw that." The class agrees; some students try to draw the hypercube, and others contend that the exponent tells how many dimensions a figure has.

After this exchange, this teacher moves on and challenges his students to investigate how they might rewrite an expression that has a negative exponent with an equivalent expression that has a positive exponent. This teacher also tells his students that they must use mathematical properties to justify their work.

### *Student A*

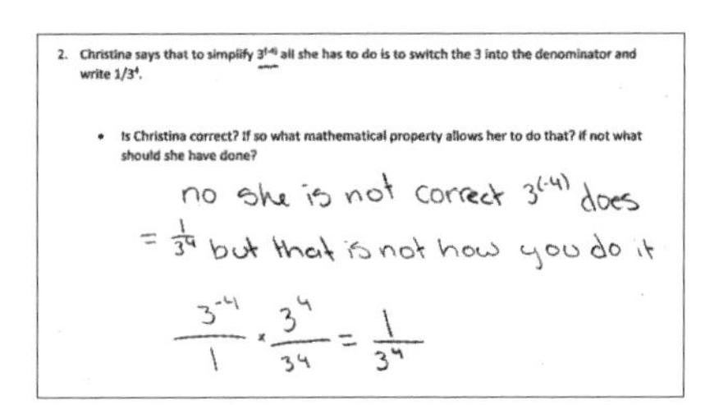


This student work demonstrates an understanding of how the multiplicative identity is applied to writing equivalent expressions for the given expression $3^{-4}$. This student also calls attention to the fact that "just switching" the integer from the numerator to the denominator neglects to represent the mathematics that actually occurs and allows "the switch" to occur.

### *Student B*

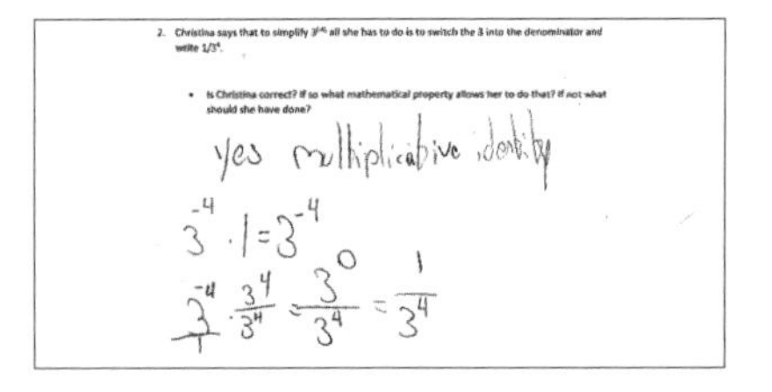


This student work clearly demonstrates an understanding that the identity property can take on many looks. The work also illustrates an understanding that any integer raised to the zero power (except zero itself) equals one. The work depicts a deep understanding of how to apply the multiplicative identity with negative exponents.

### *Student C*

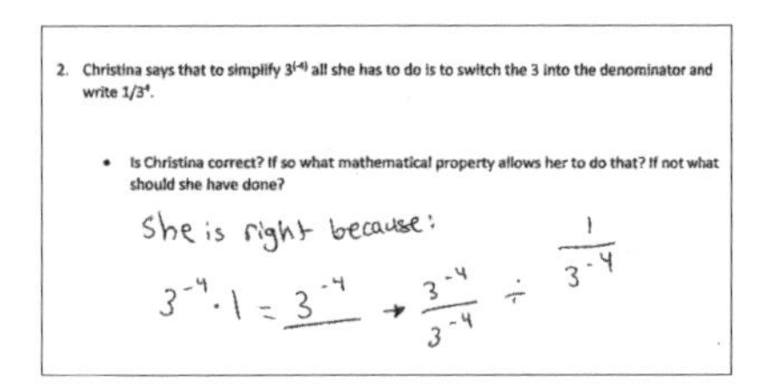


This student work illustrates some understanding of how the multiplicative identity looks. However, the work also demonstrates a lack of understanding of how to utilize that property in an application. The student switches from multiplication to division and simply moves the original expression from the numerator to the denominator. There is no evidence that the student understands that he or she is using the multiplicative identity or why. This student would benefit from having time to work with and understand the properties for working with positive, whole number exponents before introducing negative integer exponents. Modeling the rules for working with exponents using the Algeblock factor track and grid would be helpful. Once the student demonstrates proficiency with modeling situations and writing the rules he or she discovers, then this problem might be reintroduced. (There is a limit to how many dimensions can be modeled using manipulatives, though.)

## Meeting Individual Needs

Some students need a challenge, and one way to meet those students' needs is to assign the *Powerful Integers* from page A1-27 in the appendix. The problems on this reproducible challenge students to write informal proofs about why the procedures for operating with exponents work the way they do.

## Additional Reading/Resources

Lappan, Glenda, James Fey, William Fitzgerald, Susan Friel, and Elizabeth Divanis Phillips. 2004. *Growing, Growing, Growing*. The Connected Mathematics Project 2. Saddle River, NJ: Pearson.

# 9. Absolutely!

**DOMAIN:** **Domain Expressions and Equations**

**STANDARDS:** **6.NS.7.** Understand ordering and absolute value of rational numbers.

**6.NS.7c.** Understand the absolute value of a rational number as its distance from 0 on the number line; interpret absolute value as magnitude for a positive or negative quantity in a real-world situation.

**7.EE.2.** Solve real-life and mathematical problems using numerical and algebraic expressions and equations.

## Potential Challenges and Misconceptions

Working with absolute value expressions numerically, in tables, and on number lines can be confusing to students. This confusion arises from previous work in which students determine that $^{+}5$ is greater than $^{-}5$; now they must develop an understanding that the absolute value of $^{+}5$ and $^{-}5$ are the same. Both indicate 5 units from the starting point, whatever that starting point might be, as it may not always be 0. For example, on a number line 5 units from the starting point of 6 might be 11 or it might be 1, depending on the direction. Many students do not realize that the positive and negative signs represents direction and the absolute value of the integer tells the distance to zero, both of which are nonnegative. Using contextual situations involving absolute value is effective in helping students develop an understanding of this often misunderstood concept.

## In the Classroom

In one classroom, students have had many experiences walking the number line to model positive and negative integers. This teacher finds modeling on the number line so effective that she uses the same strategy to introduce *absolute value*. She directs students to line their desks around the perimeter of the room so they can all see and have easy access to the giant, floor-sized number line. She begins the lesson by asking two volunteers to go to the number line and stand at 0. She tells the first volunteer to model $^{-}7$ and the second student to move to 2. She asks the class to determine which value is greater, $^{-}7$ or 2. After allotting a minute for them to decide, she calls on a student to share an answer. Morgan volunteers, "2 is greater than $^{-}7$."

This teacher asks if anyone disagrees. When no one does, she asks her students which distance from 0 is greater, $^{-}7$ or 2. She overhears her students discussing, "Well, how can $^{-}7$ be greater? It is negative and we learned that any positive number is greater than any negative." "I can see that $^{-}7$ is farther away from 0 than 2." "I wonder if the distance should be used instead of the sign of the integer to decide."

This teacher records her students' thinking on the conjecture board before introducing the term and symbol for absolute value. She validates the student reasoning that $^{-}7$ is farther away from 0 than 2 and shows them that the symbol for denoting distance from 0 is the absolute value symbol $|\ |$. She explains that the absolute value of $^{-}7$ is represented as $|^{-}7|$ and that it denotes the distance from negative 7 to zero, which can never be negative. She then queries, "How do you think 2 is represented in absolute value notation?" She records Santana's response as $|2|$ and compliments him on his response.

She then challenges her students to make a list of things that are never negative. After a brief time, her students list distance, volume, time, money, and speed. This teacher asks whether anyone has a negation or counterexamples. Hannah timidly suggests, "Money can be negative if you owe it to someone or if your bank account is overdrawn."

Arlan asks if they can keep money on the list. He explains, "Money itself is positive; you don't have any negative coins or bills." The class agrees to think about money some more.

Next, this teacher asks a student volunteer to stand at $-6$ on the number line. Then she asks the student to move 4 spaces away from $-6$. The student walks to the left 4 units to stop on $-10$. The student tells the teacher to record the move as $-6 + 4 = -10$, which she does. The teacher asks her students if everyone agrees or if anyone has a different equation. Emmalyse volunteers, "This is wrong 'cuz if you are moving to the left you are going $-4$ units so you need to revise the equation to say $-6 + -4 = -10$." Again the teacher asks the class to discuss Emmalyse's suggestion with their shoulder partners.

Dillon and Gabe almost jump out of their seats to share their conversation. Dillon states, "The reason the first equation is wrong is because the absolute value tells us distance, so if you add $-6$ to the absolute value of 4 or $-4$ you are still going to land on $-2$."

Morgan asks if she can model what Dillon was saying. She proceeds to $-6$ on the number line and walks to the right 4 units to stop on $-2$. She tells the teacher to record the move as $-6 + |4| = -2$ and also to write $-6 + |-4| = -2$.

The teacher challenges her students to determine whether and how these two statements, $-6 + |4| = -2$ and $-6 + |-4| = -2$, can be true. She tells her students to discuss these statements with their shoulder partners. After an allotted time, she asks volunteers to share their thinking.

Angel shares his reasoning, "Because you didn't tell us what direction to move, we think both answers are correct." Frankie states, "$-6 + 4$ is not equal to $-10$, so that is wrong," Maddie adds, "Well, we think that because you told them to move 4 spaces you were talking about distance. Even though we end up in a negative space, the distance is positive." Glenn states that, "We made our own number line and showed the $-6$ where we started and what happened when we moved 4 units in both directions. We took the absolute value of $-4$ and 4 to show the distance."

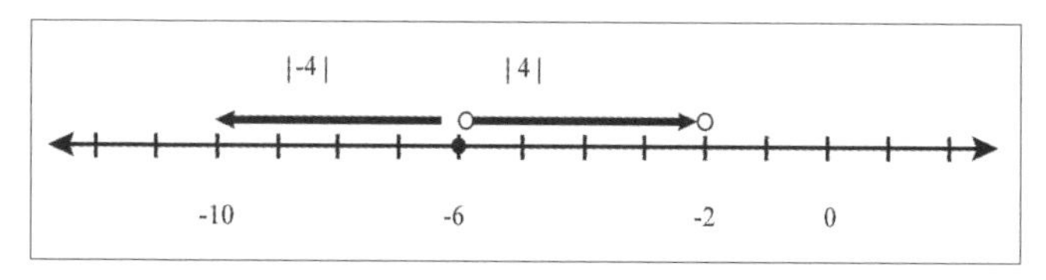


After discussing the model that Glenn and his partner shared, this teacher distributes the cards from the reproducible *Where Do I Go?* on page A1-29 in the appendix. Students are instructed to work in pairs to model each situation on a number line. After an allotted time, this teacher invites different pairs of volunteers to share their number lines. She wants all her students to realize there are multiple ways of representing the same situation or interpreting the problem. After many different methods, strategies, and interpretations, this teacher then emphasizes those that are mathematically sound and instructs her students to record those representations into their notebooks. Finally, this teacher assigned the reproducible *Absolutely!* on page A1-30 in the appendix.

SECTION 1: EXPRESSIONS

## Meeting Individual Needs

For students who need more support, tape a number line on their desk so they might model expressions with absolute value as long as they need. You may even provide them with counters so they can manipulate the results of the movement when working with absolute value. This hands-on work is often the key to understanding.

## Additional Reading/Resources


Kidd, Margaret. 2007. "Be Resolute About Absolute Value." *Teaching Mathematics in the Middle School* 12 (9): 518–523.

Wade, Angela. 2012. "Teaching Absolute Value Meaningfully: Activity Sheet for Further Exploration." *Mathematics Teacher* 106 (3).

# 10. Square Root Approximations

**DOMAIN: Domain Expressions and Equations**

**STANDARD: 8.EE.2.** Use square root and cube root symbols to represent solutions to equations of the form $x^2 = p$ and $x^3 = p$, where $p$ is a positive rational number. Evaluate square roots of small perfect squares and cube roots of small perfect cubes. Know that $\sqrt{2}$ is irrational.



## Potential Challenges and Misconceptions

Many students are so dependent on their calculators that they don't realize how much they can and should do without them. This reliance on the calculator has prevented many students from being able to calculate an approximation for the square root of larger integers and causes them to give up. And yet, it is very appropriate in many situations to estimate or approximate the value for the square root.

It is likely that some students will have had practice reducing square roots of integers via their prime factorization, but approximating square roots is also important. Approximating square roots gives students opportunities to think about the inverse relationship between squares and square roots.

## In the Classroom

This module uses some algebraic ideas to investigate approximation methods with which students might not be familiar. One, attributed to Babylonian mathematicians thousands of years ago, uses geometry that can be modeled with Algeblocks. (You might recognize a similarity with a geometric version of completing the square.) Reconnecting algebra and geometry is a useful reminder that all mathematics is related.

The class begins with one teacher asking his students to compute the exact or approximate (to two decimal places) values of the square roots of 7, 29, 30.25, and 51 *without the use of a calculator*. He distributes the *Broken Calculator* reproducible from page A1-31 in the appendix and directs small groups of two or three students to work on the four computations. In one group, Peter is heard explaining to his partners that "we need to find the positive number whose square is 7. Because 7 is between 4 and 9, I know the square root of 7 is between 2 and 3. Also, because 7 is closer to 9 than it is to 4, I'm going to guess that the square root of 7 is greater than 2.5."

Marcia agrees and adds, "Right; 2 squared is 4 and 3 squared is 9, so the square root of 7 has to be between 2 and 3. And I know that 25 squared is 625, so 2.5 squared has to be 6.25. That is close to 7 but is still too small."

Jacqueline agrees and offers, "Well, that's true but 6.25 is closer to 4, only 2.25 away, than it is to 9, which is 2.75 away. I'm going to try 2.65 next." She multiplies on paper for a few minutes. "Well, 2.65 × 2.65 is 7.022 and 2.64 × 2.64 is 6.9696. I say the square root of 7 is between 2.64 and 2.65. I'm going to say it's approximately 2.65, to two decimal places."

This teacher asks whether anyone got a different answer or did it differently.

Jackson replies, "I drew a square of 2 × 2 and another one 3 × 3, but I don't know what else to do. All this does is to show that the square root of 7 is between 4 and 9." The class agrees that Marcia's way makes sense but is a lot of work, and they want to know if there is an easier way.

This teacher introduces the Babylonian square roots method to his students. He tries to incorporate some history of mathematics whenever he can. The Babylonian square roots method takes students through a geometric method, developed thousands of years ago in Mesopotamia (present-day Iraq), that can be used to approximate square roots. This historical method engages students with SMP 7: Look for and make use of structure.

This teacher provides his students with Algeblocks and asks them to model this method as they follow along with his demonstration. He insists that they draw a pictorial representation as well. He begins by using the method to approximate $\sqrt{27}$.

Because $\sqrt{27}$ is the side length of a square having area 27, he tells his students to try to build a square with area 27 square units. He asks his students what the closest dimensions might be. Jillian suggests they start with the reasonably close approximation of 5. This teacher agrees and then tells his students to model the square using the blocks and the factor track. He also tells them to assign a value of 5 to each of the yellow rectangles and to fill in the space created by the two factors with the yellow square tile. He asks them to tell him what they notice. Jack points out that the 5-by-5 square has an area of 25, so it is a square that is the right shape but has the wrong area.

The area is off by 2, so Jack suggests they can add two rectangles, each having an area of 1 (and one side length of 5), to the side of the 5-by-5 square to create a rectangle with area 27.

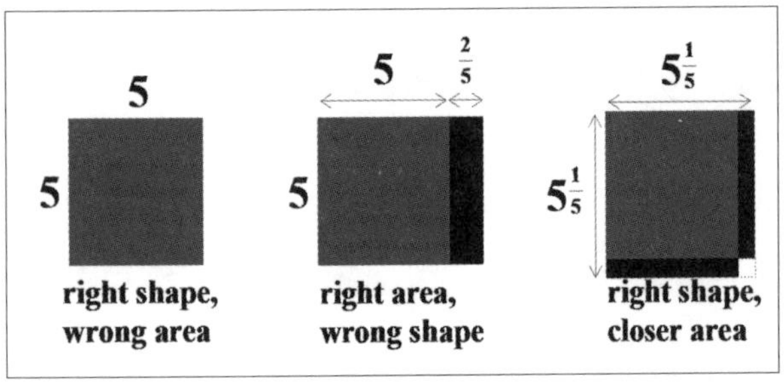

5 5 right shape, wrong area | 5 $\frac{2}{5}$ 5 right area, wrong shape | $5\frac{1}{5}$ $5\frac{1}{5}$ right shape, closer area

This teacher explains that for the two rectangles to have a total area of 2, each rectangle must have an area of 1. This means that the dimensions of each new rectangle is $5 \times \frac{1}{5}$. He adds rectangles as Jack suggests.

Again, he asks what the students notice. Spencer states, "Now we have the correct area, but it's the wrong shape. It's not a square any longer."

This teacher explains that if he moves one of the rectangles to the bottom of the square, the new shape is *almost* a square. (Notice that it's actually a $5\frac{1}{5} \times 5\frac{1}{5}$ square with a tiny $\frac{1}{5} \times \frac{1}{5}$ square removed.) The Babylonian approximation for $\sqrt{27}$ is therefore $5\frac{1}{5}$, or 5.2. After asking if there are any questions, this teacher assigns the Babylonian Square Roots reproducible on page A1-32 in the appendix.

This teacher also shares one other geometric model. He distributes graph paper and instructs his students to draw a square that has an area close to 27 square units. He invites Jazz to the board to do her work. Jazz outlines a 5-by-5 square on the graph paper. Next, this teacher tells the class to outline a 6-by-6 square and to ensure that the 5-by-5 square is inscribed within it. He asks Leonard how many additional square units he added. Leonard shares his squares and explains that he had 25 square units, added 11 square units, and now has a total of 36 square units. This teacher asks Leonard how many of the additional 11 squares he needs to make an area of 27 square units. Leonard responds that he needs 2 of the 11, or $\frac{2}{11}$. This teacher finally asks his students what the side length of the new square will be. In unison, the class calls out $5\frac{2}{11}$. He provides the decimal equivalent to three decimal places, 5.182.

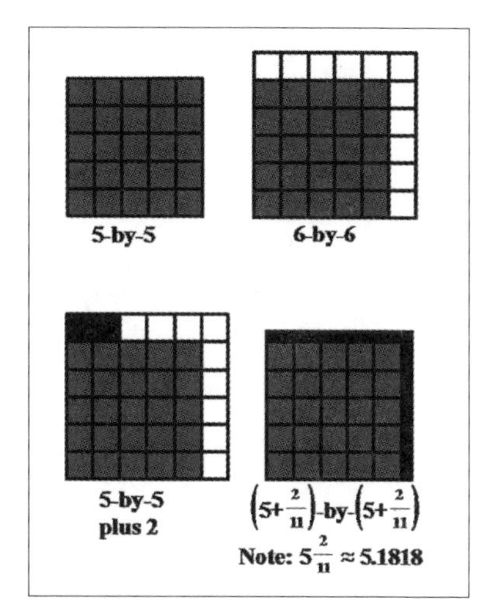


Although not as exact as the Babylonian method, this model will also render a close approximation for a square root of a given number. Notice that this method gives an approximation that is too small, and the Babylonian Method gives an approximation that is too large.

## Meeting Individual Needs

Many students may need the support of graph paper and colored pencils to sketch the two squares that border the radical for which they are trying to find the root for a longer time than others. This is fine because it does provide a visual representation of how square roots work, and it links the vocabulary of "square root" to actual squares.

The challenge problem from *Babylonian Square Roots* and the entire *Try This!* activity can be used with more advanced students. *Try This!*, page A1-33 of the appendix, has students work through the details of a square root approximation method they may have seen in books (it used to be in the *World Book Encyclopedia*) or on the Internet.

## Additional Reading/Resources

Flores, Alfinio. 2014. "The Babylonian Method for Approximating Square Roots: Why Is It So Efficient?" *The Mathematics Teacher* 108 (3): 230–235.

Wiesman, Jeff L. 2015. "Enhancing Students' Understanding of Square Roots." *Teaching Mathematics in the Middle School* 20 (9): 556–558.

# Equations

## 11. Activity Centers

**DOMAIN: Expressions and Equations**

**STANDARD: 6.EE.C.** Represent and analyze quantitative relationships between dependent and independent variables.

### Potential Challenges and Misconceptions

All students need to recognize that there are linear and nonlinear relationships and that they can be represented in tables, graphs, expressions, and equations. In addition, it is important for students to distinguish between continuous and discrete data as represented in tables and graphs. Many students struggle with these concepts because most or all their elementary experience has revolved around bar graphs, and if they did draw a linear graph, they always connected the points. To connect points on a graph or not is a new consideration for most students. All students benefit from varied experiences in contextual situations as they meet the challenge of determining whether points are connected or not. For example, it is not probable to have half a marble, but it is probable to have half a cup of milk. Students also need multiple experiences determining how the data in a table look in a graph and vice versa. The inclusion of activity centers in the algebra classroom is an effective way to challenge, remediate, or meet the needs of grade-level students.

### In the Classroom

One teacher projects *What Kind of Graph Am I?* from page A2-2 in the appendix, and distributes the reproducible to his students. He instructs his students to examine each graph and determine whether the graph is linear, quadratic, exponential, absolute value, or something else. He encourages his students to collaborate on the identification process and instructs them to record the reason for their selections. These activities engage students in several of the Standards for Mathematical Practice.

Prior to class, the teacher has set up a variety of activity centers. The centers present activities that review a previous model for representing expressions or represent a different relationship and result in a different graphical representation. At each center, students work in small groups to record the data from their activities on easel-sized paper and make their own conclusions about the shape of the graphs. This teacher allows time for his students to complete at least two activities per class period, with all four completed in two days. After all students have completed tables and graphs on easel-sized paper for each station, this teacher facilitates a gallery walk. He tells his students to match the graphs to the corresponding activity.

### *Activity Center One*

This center has five vases of various shapes and sizes, rulers, a pitcher with colored water (or rice), a small scoop, and paper towels. Students are required to fill each vase with scoops of colored water and record the number of scoops and the height of the water in the vase after each scoop of water is added. The resultant graphs are typically nonlinear, unless the vase is a cylinder, with the graphs taking on the shape of the vase.

### *Activity Center Two*

This center has a set of cards on the table. Students are directed to match the graph with its corresponding description See *What Kind of Graph Am I?* from page A2-2 in the appendix.

### *Activity Center Three*

This center will focus on building factor lattices using variables. The center has a box of either gumdrops or dots candy (small Styrofoam pieces or play dough also work) and toothpicks. Students make a factor lattice for $xy^3$. Directions are on *Factor Lattices* 2, page A2-3 in the appendix. Students use the factor lattice to determine the greatest common factor and least common multiple between various algebraic expressions. Students will also represent the expressions using Venn diagrams as an alternative representation.



### *Activity Center Four*

This center focuses on modeling exponential growth situations. The center has a stack of paper and scissors on the table. The directions for this center instruct students to take one sheet of paper, cut it in half, and record in a table the number of cuts and number of resulting pieces of paper. Next they stack the pieces and cut the stack in half. They repeat this cutting process until they cannot cut the paper again. Students are directed to make a table to show the relationship between the number of cuts and the number of pieces of paper that result. They graph the data in the table and write an algebraic expression that best models the relationship.

### *Activity Center Five*

This center has a container of Algeblocks with instructions to model a variety of expanded and factored representations. Students are told to build a model for each expression, sketch the model, and write an algebraic expression that best represents the results of the operations performed in the model. Sample expressions can be found in *Modeling Algebraic Expressions* on page A1-22 in the appendix.

## Meeting Individual Needs

The inclusion of the activity centers allows this teacher to provide differentiated instruction to students who need more support and simultaneously provide a challenge for students who need a deeper exploration of algebraic concepts. Teachers should assign students to the appropriate centers to ensure they are being challenged, reviewing difficult concepts, or reinforcing learning that may be fragile.

## Additional Reading/Resources

Day, Lorraine. 2015. "Mathematically Rich, Investigative Tasks for Teaching Algebra." *Mathematics Teacher* 108 (7): 512–518.

Polygraph by Desmos. https://teacher.desmos.com/polygraph.

SECTION 2: EQUATIONS

# 12. Balance Beams

**DOMAIN:** **Expressions and Equations**

**STANDARD:** **6.EE.5.** Understand solving an equation or inequality as a process of answering a question: which values from a specified set, if any, make the equation or inequality true?

## Potential Challenges and Misconceptions

Many students know procedures for solving equations but have yet to internalize that the equal sign actually means a "balance" between two quantities. Because most students experience the equal sign as meaning "get the answer," when challenged to solve an equation that does not require the final answer, they display confusion. Many students also miss the concept that when adding quantities, the number represents a quantity and is an adjective telling how many of something is being described. The noun that the number describes must be the same if the quantities are combined. For example, 1 + 1 = 6 makes no sense unless there is a context such as 1 penny + 1 nickel = 6 cents or unless it is made known that both quantities are describing the same object.

SECTION 2: EQUATIONS

## In the Classroom

To assess her students' prior understanding of solving equations, one teacher assigned her sixth- and seventh-grade students the following equation.

$$4 + 3 = \square + 5$$

The student work reveals some major misconceptions, partial understandings, and computational errors. One common error was the inclusion of run-on equal signs and a second was the misconception that the missing value represented a final answer. Examine the following samples of student work to understand how the preponderance of students solved the problem.

### *Student A*

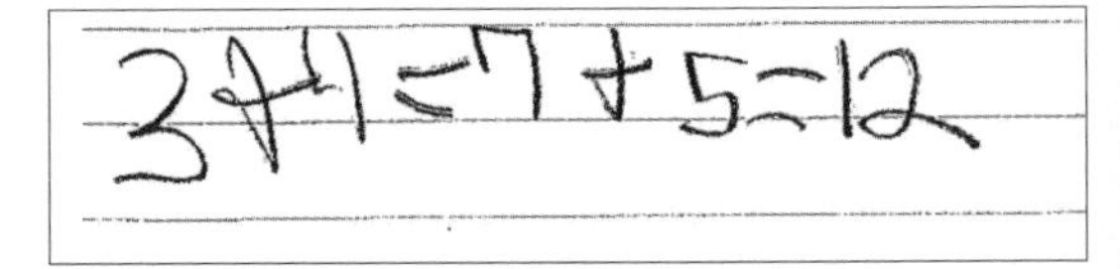


The evidence in this work illustrates the misconception that the unknown is simply the answer from the left side of the equal sign, and this student appears to ignore the fact that there are *two equivalent expressions* that must be balanced. This student also uses a second set of equal signs, neglecting to understand the transitive property of addition that states if $a + b = d$ and $b + c = d$, then $a + c = d$. In this case, 3 + 4 would have to equal 12, and that is just not the case.

### *Student B*

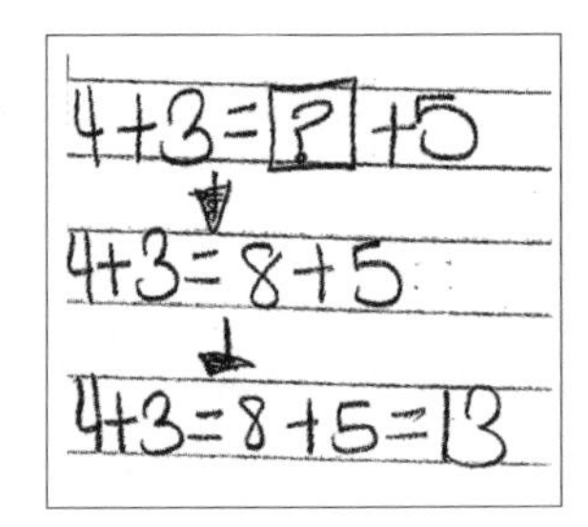


The evidence in this work also illustrates the misconception that "run-on" equal signs can be used, regardless of whether they represent true statements. It is evident that 4 + 3 does not equal 8, although it is true that 8 + 5 equals 13. Again the run-on equal signs indicate that the student uses them to list the steps that the student performed: "First I added 4 + 3, then I added 8 + 5, then I got 13." The work shows a lack of understanding about the balance property that must occur with equations.

### *Students C and D*

These students appear to have ignored the equal sign altogether—or do not understand the significance of it—and add numbers together until one number results. These students need experiences to help them understand that the equal sign indicates that two expressions are equivalent. The evidence in this work points to serious misconceptions and the need for an immediate intervention using manipulatives so that the students develop concrete understanding before moving on to solving the more abstract equations. They would benefit from using counters as weights and modeling the original equation and its solution on a pan balance. It would be extremely beneficial for this teacher to interview these two students to determine how they are thinking about the equations.

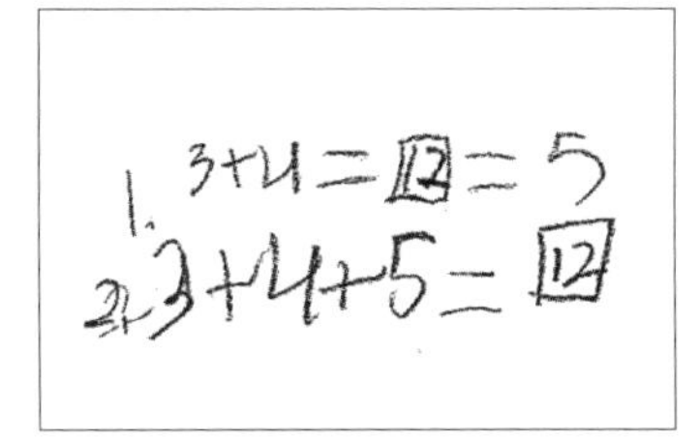


Student C

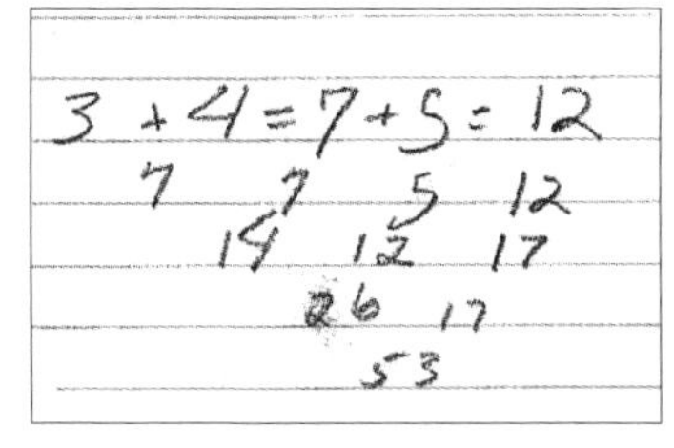


Student D

This teacher uses each of these samples as problems she assigns the class. She tells students that their job is to find the errors and discuss with a friend what is wrong and how to correct it (SMP 3—create a mathematical argument and critique the reasoning of others). Jamal's group is excited to share its conversation with the class. He announces, "We figured out that we can use our thumbs to cover parts of the equation to test to see if our answers are correct. So, if you look at 3 + 4 = 7 + 5 = 12 put your thumb on the 7 + 5 to hide it and see if 3 + 4 = 12. See, it doesn't, so the answer is wrong."

Leila's group approaches the problem differently. "Because the original problem is 3 + 4 = *some number* + 5, there shouldn't be another equal sign. What you need to find out is what that number is—what number + 5 is equal to the same amount as 4 + 3." This teacher then assigns the *Balance Beams* reproducible from page A2-4 in the appendix.

Once the students understand the mathematical notation for writing a balanced equation, this teacher, with an eye toward helping her students develop a conceptual foundation for combining like terms, asks whether 1 + 1 = 6. Immediately the students call out, "No way!" This teacher reacts by asking, "What is the total of one nickel and one penny?" A chorus of "six cents" is heard in the classroom. This teacher agrees and comments that it is very important to think of the context in which numbers are given, and if a context is missing the convention is to assume the same objects are being combined. She then challenges her students to complete the reproducible *Can These Really Be True?* on page A2-5 in the appendix. This activity is very helpful in connecting measurement conversions, equations, and the importance of context.

## Meeting Individual Needs

One teacher provides pairs of students with balance beams and allots time for them to experiment with various equations. For example, on the left side of the fulcrum, students might place three weights at the number 2. This actually represents the value of 6 (3 × 2). On the right side of the fulcrum, they should be encouraged to find as many equivalent values as possible. They may place one weight on the 6, or one on the 4 and one on the 2, or two weights on the 3. Students should record each equation either in their notebooks or on a recording sheet. To ensure random equations, she tells her students to toss two dice and use those values on one side of the fulcrum. Once those values are set, she tells them to roll one die to determine the value to be built on the other side of the fulcrum. She then tells her students to determine the missing value that will balance the scale.

If balance beams are not available, you may use pan balances or scales that are available in most science classrooms. There are also several virtual balance manipulatives available online; the Balance Beam on the National Council of Teachers of Mathematics' Illuminations website (https://illuminations.nctm.org/) and the Balance Scales on the National Library of Virtual Manipulatives website (http://nlvm.usu.edu/en/nav/vlibrary.html) are two.



## Additional Reading/Resources

Szydlik, Jennifer Earles. 2015. "Mathematical Conversations to Transform Algebra Class." *The Mathematics Teacher* 108 (9): 656–661.

# 13. Challenge and Intervention Centers

**DOMAIN:** **Expressions and Equations**

**STANDARD:** **Standard: 6.EE.5.** Understand solving an equation or inequality as a process of answering a question: which values from a specified set, if any, make the equation or inequality true?

## Potential Challenges and Misconceptions

Algebraic manipulation for the sake of manipulation too often stunts students' ability to progress in their algebraic thinking and reasoning. When students have an opportunity to "play" with equations using mobile balances, puzzles, and games, they have better access and opportunity to see why what is done to one side of an equation must be done to the other side. Most students also respond positively to the challenge of making their own balance mobile problems.

## In the Classroom

In many classrooms, students are heterogeneously grouped; students with special needs and students needing greater challenge are mixed with regular education students. One teacher addresses the diversity in his classroom by using activity centers. When he is working with his students solving equations, he sets up five activity centers through which students rotate and solve a variety of equations of varying complexity. He makes it clear that the goal is not to hurry through each of the centers but, rather, to think and reason through the problems and to model various representations. He instructs students to ask guiding questions when peers are struggling, rather than showing or telling them answers.

This teacher also names each of the centers with a noted mathematician such as Carl Friedrich Gauss, Ada Lovelace, Leonhard Euler, Sophie Germain, Hypatia, and Al-Khwarizmi. He places placards on the center table with a brief biography. He thinks this subtle explanation of famous mathematicians may influence some students. (See *Mathematicians* on page A2-6 in the appendix.)

At the Gauss center, this teacher provides a series of problems related to arithmetic series and counting a variety of rectangular arrays. This teacher places tiles, Cuisenaire rods, and graph paper to encourage his students to model the problems they are trying to solve. The Gauss activity challenges students to determine the equation for Gauss's theorem through investigation. Sample student work illustrates the range of understanding of how Gauss's theorem might be represented and how various students think about writing equations. Because most students think that all the work they do on a given day follows the same pattern, the reproducible also includes a problem that represents the sum of the square numbers. See *Let Me Count the Ways* and *Gauss Center Activity Cards* from pages A2-8 to A2-9 in the appendix.

At the Ada Lovelace center, this teacher addresses the concept of keeping a balance while solving equations represented as mobiles. This fun but challenging activity has proven successful for engaging middle school students in solving equations and providing a review for high school students. This teacher scaffolds the puzzles to provide access and challenge for all his students. See the *Mobile Madness* reproducible on page A2-10 of the appendix for example puzzles.

This teacher also challenges students to build their own mobiles. He provides mobile frames that he has collected over the years from yard sales, as well as coat hangers, string, cardboard squares, and felt markers. He also places pattern blocks and sticky circles on which students can record numerical values. For the pattern block mobiles, students explore the mathematical relationships among the pattern blocks. For example, three triangles weigh the same as a trapezoid, six triangles weigh the same as a hexagon, three rhombi weigh the same as the hexagon, and two trapezoids weigh the same as a hexagon. On index cards, this teacher has drawn in black marker the shape of a mobile and challenges the students to place the appropriate numerical values where they will balance on a mobile. Some samples are on the *Pattern Block Mobiles* reproducible, page A2-11 in the appendix.

The Euler center poses the classic Knight's Move challenge. This teacher provides chessboards complete with plastic chess knight pieces to allow his students to model the activity. He suggests that students try the problem individually before collaborating with a partner. He also places a number of blank chess grids at the center, as this activity may begin with a guess-and-check strategy. The activity can be found on The *Knight's Move Challenge*, page A2-12 in the appendix.

At the Sophie Germain center, this teacher places Cartesian coordinate plane sets of puzzles for students to graph. Each set of ordered pairs offers practice graphing in the four quadrants. Each set of ordered pairs, when graphed, results in a picture. One example is *Something Is Fishy*, page A2-14 in the appendix. The Sophie Germain center also includes *Shape Equations*, page A2-16 in the appendix, which engages students in solving equations using various shapes as variables.

At the Hypatia center, this teacher provides cards with linear and nonlinear patterns. The challenge for students is to complete the patterns and explain in words or with an expression how the pattern grows. The *Hypatia Center Activity Cards* are on page A2-17 in the appendix.

At the Al-Khwarizmi center, students match an inequality statement to a number line and context. The set of cards can be found on the *Al-Khwarizmi Activity Center Cards* reproducible, page A2-18 in the appendix. To support the fact that multiple values can be the solutions for inequalities, this teacher places pan balances and blocks at the station for those students to experiment with various solution possibilities. This teacher also plans time after all students have completed the center activities to discuss how context can affect the graphs of solutions for inequalities.



## Meeting Individual Needs

Encourage students who need more scaffolding to use balance scales to model the puzzles and see the results of what happens if they do not maintain equivalent values on both sides of the fulcrum. Most students who understand how equations operate also respond positively to the challenge of writing balance mobile problems. Writing a problem situation requires greater cognitive demand than just solving a problem. It is also fun to include Ken-Ken puzzles and Sudoku puzzles for students looking for more activities.

## Additional Reading/Resources

Snapp, Robert R., and Maureen D. Neumann. 2015. "An Amazing Algorithm." *Teaching Mathematics in the Middle School* 20 (9): 540–547.

# 14. Gauss and Figurate Numbers

**DOMAIN:** **Expressions and Equations**

**STANDARD:** **HSA.CED.1.** Create equations and inequalities in one variable and use them to solve problems. *Include equations arising from linear and quadratic functions, and simple rational and exponential functions.*

## Potential Challenges and Misconceptions

Recognizing and writing quadratic equations is truly a challenge for middle school students who are accustomed to only working with linear equations. However, when students are presented with opportunities to model and draw pictorial representations of quadratic situations, the transition to working with quadratics is much less threatening. It is important that all students recognize that quadratics are based on squares, are written with an exponent of two, and when graphed have a maximum or minimum vertex.

## In the Classroom

This teacher begins her lesson on counting by asking the students to stand and take turns giving each classmate a fist bump. She purposely neglects to tell her students to count how many fist bumps are exchanged; rather she wants them to experience different counting situations. After the students complete the fist bump with each other, this teacher divides the class into halves. This time she asks each student from one half to give a fist bump to each student in the other half. Next, she asks her students to share with a shoulder partner how those two situations are the same and how they are different. She also asks them to predict which context will result in the most fist bumps exchanged. (In the past when she did these types of problems, she did not include her students acting out the situations and discovered many were confused with the differences between them.)

After the quick activity, this teacher projects this problem:

> *There are twenty-eight students in our math class. After every student competed in the Math Counts contest, each student gave every member in the class a fist bump. How many fist bumps were exchanged? List and show what strategy or strategies you might use to solve the problem. Justify your response.*

This teacher asks her students to work individually on the problem for few minutes, after which she calls the class back to attention. She asks students to share the strategies they are thinking about, which she records on the board. She hopes that if she lists all the strategies students are thinking about, those students who are unable to define a strategy will be able to choose among those on the board. This facilitated guidance gives all students access into the problem and supports the importance of using a problem-solving strategy rather than just jumping into computing an answer. The list of strategies includes:

- Solve a simpler problem.
- Act it out.
- Make an organized list.
- Draw a picture.

SECTION 2: EQUATIONS

After sharing the strategies, this teacher instructs her students to solve the problem. The following student work illustrates the various strategies and representations students used.

### Student A

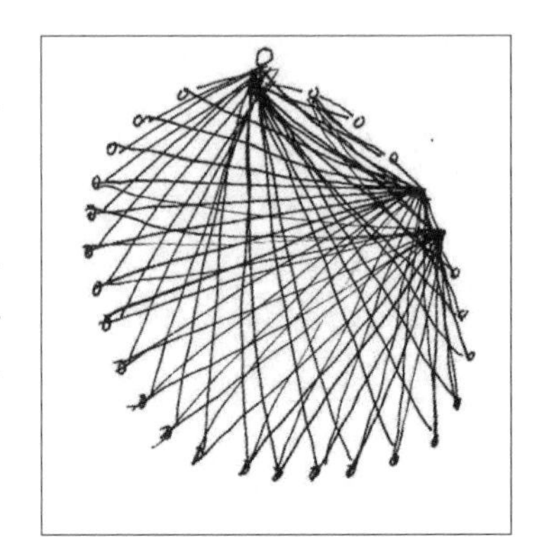

This student attempts to draw a picture but quickly realizes it is going to be difficult to count the number of fist bumps. This teacher asks Student A to hang the picture on the board and to compare it with the work submitted by Student B. Student A tries valiantly to work with all twenty-eight fist bumps at once. This student would benefit from recognizing the value in solving a simpler problem, in this case, starting with one person and building up to the point where a pattern emerges.

### Student B

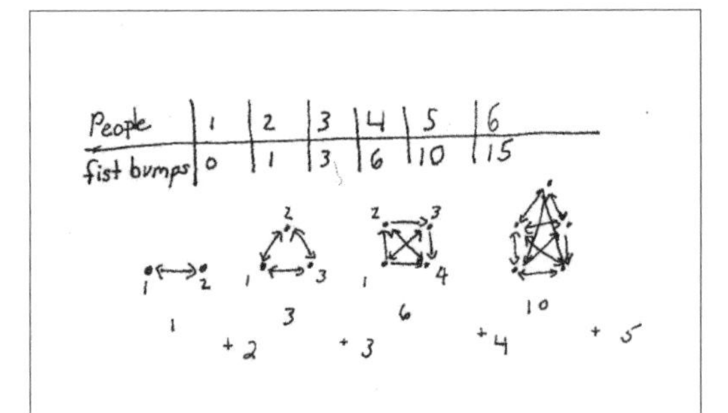


This student chooses to make a table to complement a geometric representation of how it looks to give a fist bump to a variety of numbers of teammates. This student both draws a geometric representation and makes a table. This student solves a series of simpler problems, beginning with one student and working up to six students, when a pattern emerges. The evidence in the work indicates the student found a pattern that shows the number of fist bumps is increasing by the consecutive integers. This student is unable to find a general equation but illustrates an understanding of how valuable it is to use a problem-solving strategy. What is more important is the logical, sequential manner of representing the number of fist bumps together with the geometric representation. Notice that as the geometric shapes develop, they get almost as messy as Student A's drawing.

### Student C

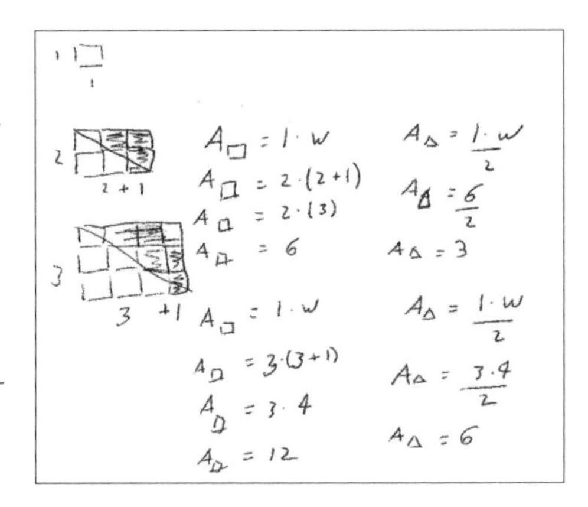


The evidence in this student work shows that the student solves a simpler problem using an area model. The student starts with one square but does nothing with it. With the second level, he or she draws a right triangle then adds an additional right triangle to make a rectangle. He or she finds the area of the rectangle and then halves it to find the area of the triangle. This student work appears to relate back to a prior experience this student has had. It is possible that this student has had experience using Cuisenaire rods to model some counting problems. Because there is a lot of ambiguity about how or why this student did what he or she did, it is important for this teacher to interview the student.

### *Student D*

After seeing the work shared by Student B, Student D takes time to generalize the equation. This student is able to show that $\frac{2(2+1)}{2}$ and $\frac{3(3+1)}{2}$ would generalize to the expression $\frac{n(n+1)}{2}$. This student also substitutes in values to support his or her conjecture.

This teacher realizes that students are capable of building on one another's ideas, as Student D demonstrates with the formula for finding the sum of consecutive integers. She also notices that Student D is also thinking about SMP 7: Look for and make use of structure. This teacher recognizes the importance of posting student work for the whole class to examine, critique, and form mathematical arguments.

Upon completion of this problem and the summary, this teacher assigns *Let Me Count the Ways* from page A2-8 in the appendix. This activity challenges students to count in a variety of contexts and to demonstrate their proficiency in identifying and using a variety of problem-solving strategies.

## Meeting Individual Needs

For students who need more scaffolding, it is helpful to work with the students to solve simpler problems—not similar problems—by helping them understand it is advantageous to break the problem down into smaller parts. For example, if students are trying to determine how many tetrahedra there are in a tetrahedron with thirteen rows, they might start with how many tetrahedron there are in one row, then in one with two rows followed by three rows, and so on. (It is also fun for students to actually build the tetrahedron using drinking straws.) By recording the results in a table, students are able to see patterns, and the table helps identify the expression that best represents the situation. This is a strategy that is often overlooked but often leads to the general formula for many situations.

## Additional Reading/Resources

Chan, Helen Hsu. 2015. "How Do They Grow?" *Teaching Mathematics in the Middle School* 20 (9): 820–824.

Kenney, Margaret J., and Stanley Bezuszka, S.J. 2001. *The Number Treasury*. Palo Alto, CA: Dale Seymour Publications.

Szetela, Walter. 1999. "Triangular Numbers in Problem Solving." *The Mathematics Teacher* 92 (9): 820–824.

# 15. Equations in All Forms

**DOMAIN: Expressions and Equations**

**STANDARD: 8.EE.6.** Use similar triangles to explain why the slope *m* is the same between any two distinct points on a nonvertical line in the coordinate plane.

## Potential Challenges and Misconceptions

Many students are introduced to linear equations with no prior experience with graphing ratios. Because the slope of a line is a ratio, it is important for all students to have experience graphing ratios, comparing multiple ratios on the Cartesian coordinate plane, and seeing that equivalent ratios have the same slope. Also important is for them to view coordinate pairs as points on a line and to recognize that different forms of equations can be graphed. Students who have multiple experiences with situations in which they have, or can identify, a point and a slope, two points, or an equation in standard form are better able to move interchangeably among the different representations. Teaching each in an isolated fashion prohibits many students from gaining the flexibility they need to respond to situations given various data representations.

## In the Classroom

In one classroom, this teacher begins her introduction of ratios as relationships that can be graphed on the Cartesian coordinate plane. She brings her classes to the nearest stairwell and asks students to measure the height and depth of each stair. She asks them to label the height of one step as the rise and the depth as the run. She also has them measure the height of the banister, from top of a stair to the top of the banister, and the total length of the banister. She provides triads with meter sticks and measuring paper. She suggests that some of the students begin at the top of the stair set looking down and others at the bottom of the stair set looking up. When the data are collected, the students return to their classroom. Because the students have measured the same sets of stairs, this teacher asks the students to think about the data she recorded in the table.

| Rise | 6 | 5.5 | 6.26 | 5.8 |
|---|---|---|---|---|
| Run | 12 | 11.5 | 10.5 | 11.75 |

This teacher allots time for the students to make sense of their data. This short time is mainly for students to organize their data in a way in which they can explain what they discover. She begins the discussion by asking volunteers to share what they observe. Charlie comments, "Because we all measured the same stairs, then our data should be the same. I think there are mistakes in the measurements."

This teacher asks Charlie and the rest of the class what they should do about the discrepancies in the data. Henry quickly responds, "Because all the data are close to 6 inches and 12 inches, we can use those friendly numbers." The students nod their heads in agreement, but Sam is insistent he use his 6.26 and 10.5 inches.

This teacher continues the discussion by asking for other observations about the data. Yazmin begins by saying, "All the stairs in a staircase have the same height and same depth."

Jax adds, "I measured the banister and got 24 for the rise and 48 for the run. I noticed that if I made a ratio of the height to the distance from one end to another, the simplified fraction is the same as the fraction of my height to my depth of a stair . . . isn't that cool?"

Christian adds, "When I measured the banister from the bottom to the top I got one answer, but when I measured it from the top to bottom, I got a negative answer. I think it matters if you start at the top or the bottom of the stairs."

When there are no more observations, the teacher asks her students how they might represent their data. Ming suggests, "What if we made ratios? We could compare the depth to the height." With this idea proposed, the teacher wants to determine what they knew about ratios. She asks her students to make conjectures about ratios. She asks them to tell her everything they know about ratios. After two or three minutes, she records her students' responses on the conjecture board.

Conjectures

What are ratios?

- they are fractions
- they compare things
- they tell us things
- they tell us about numbers
- they tell us about parts to parts or parts to wholes

This teacher invites her students to think about the conjectures and after a minute to turn to a shoulder partner and discuss whether or not they can negate or find a counterexample for any of the conjectures. After an allotted time, she returns to the conjecture board and crosses out those ideas that do not always work mathematically, that are not specific enough to define ratios, or that students prefer to qualify or revise.

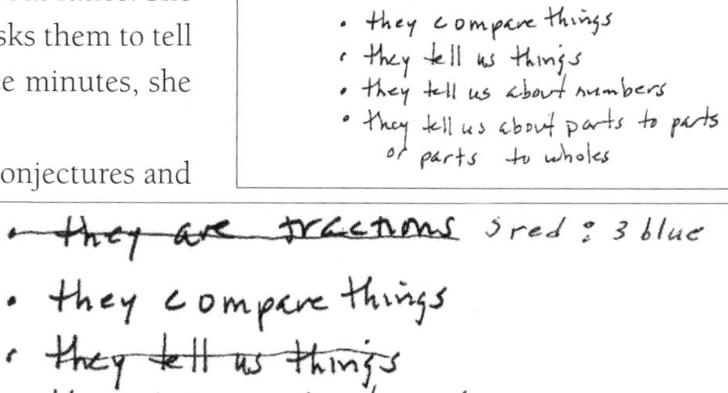


After revising the conjectures, this teacher asks her students to consider if the order in which they compare the depth and height of the stairs matters. She provides her students with time to collaborate and ensures they have graph paper as a prompt to think about the $x$- and $y$-axes. As she walks around among the groups, she suggests to some groups they graph their data; at other groups, she asks whether the height of the stairs is best represented on the $x$- or $y$-axis. After this short time, she invites students to share their thinking. Mario volunteers and suggests, "I think it is important that the height is represented on the $y$-axis and the depth on the $x$-axis. I think that because the $y$-axis shows height." The rest of the class concurs and this teacher writes the ratio as height:depth. She then states that some people think about this as the ratio of the rise to the run. At this time, she suggests her students graph their data. She invites two students to make their graphs on easel-sized paper and hangs them on the board.

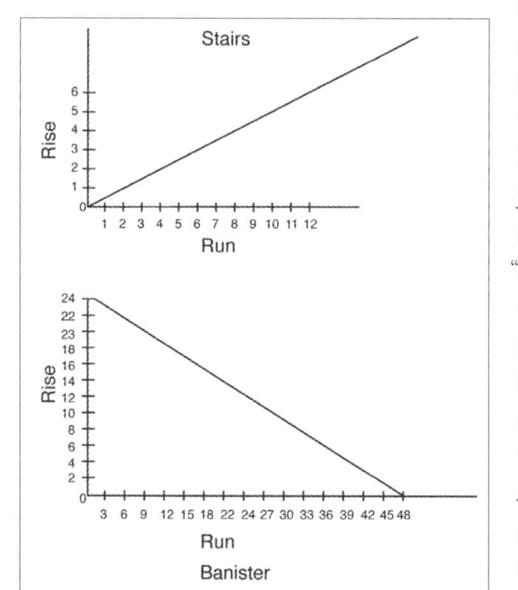

Once the graphs are displayed, this teacher instructs her students to draw line segments connecting the relationship between the rise and the run of the data. Her students confirm that the line they drew goes through the vertices of the rectangles and that the points are equivalent.

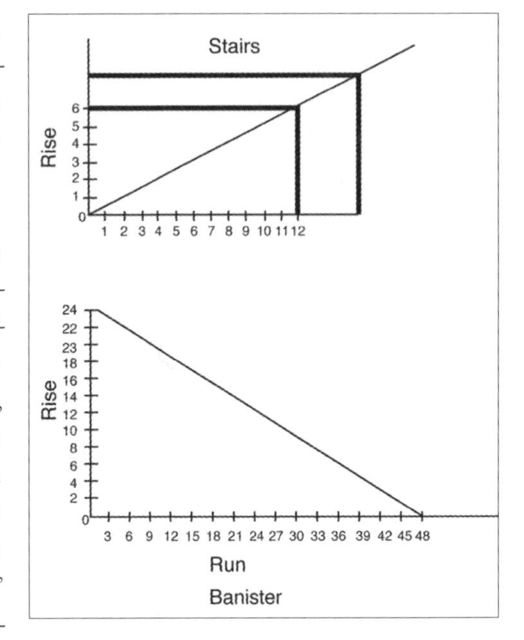


This teacher asks the students to compare the two graphs and make a list of similarities and differences. She asks them to consider how they might differentiate between the directions of the two lines. She is not surprised when Dexter, an avid skier, states, "You can tell how steep the slope is by the angle that it forms and the direction that is goes. Like in the first graph, we are at the bottom of the ski lift heading up the mountain, and on the second graph, we are at the top of the mountain getting ready to ski down."

This teacher follows this activity by asking a volunteer to stand on (6, 5) on the floor grid. She asks a second student to move to a point to create a line with a slope of $-\frac{4}{3}$. She asks two more students to stand on additional points on the grid. She also invites an additional student to move to the whiteboard and draw the graph. While the volunteers are modeling this situation, students in their seats are doing the same activity on their individual whiteboard grids. Once all the students have determined the additional points on the grids, the class discusses the results.

This teacher asks the students adding points how they know whether they should make the $y$-term increase or decrease by three or if it matters. This question often leads to students asking whether the negative sign is for the three, the four, or both. This teacher poses a question in response, "Is the numerator, denominator, or ratio negative? And for a ratio to be negative, how many terms must be negative?" The class agrees that either the numerator or the denominator can be negative, so the following points should lie on the line: (3, 9), (0, 13), (9, 1), (12, -3), among others.

After practicing a few more examples by modeling, this teacher assigns *Equations in All Forms* from page A2-19 in the appendix.

After examining the various student responses, the students agree that they should add the conjecture that *ratios represent the slope of a line* and that *all slopes are ratios*. This teacher adds these two conjectures, and then assigns the *Graphs and Equations* reproducible from page A2-20 in the appendix.

## Meeting Individual Needs

Many students benefit from more concrete modeling of generating the points on a line as they seek to write an equation given a point and a slope or two points. In addition to making human plots, it is often helpful for these students to work with the general form of an equation and to find the $x$- and y-intercepts. This can be done by covering up first the $ax$ term and finding the value for $y$. The covering up represents the $a(0)$ product to find the $y$-intercept. Next, cover up the $y$ term to find the $x$-intercept when $b(0)$. Once those two intercepts are found, you can determine the slope.

## Additional Reading/Resources

Ceyanes, Pamela Lockwood, and Kristina Gill. 2014. "Three Lessons on Parabolas—What, Where, Why." *Teaching Mathematics* 108 (5): 368–375.

Collins, Anne, and Linda Dacey. 2011. The *Xs and Whys of Algebra: Key Ideas and Common Misconceptions*. Portland, ME: Stenhouse.

# 16. Linear Systems

**DOMAIN:** **Expressions and Equations**

**STANDARDS:** **8.EE.8b.** Solve systems of two linear equations in two variables algebraically, and estimate solutions by graphing the equations. Solve simple cases by inspection.

**8.EE.8c.** Solve real-world and mathematical problems leading to two linear equations in two variables.

## Potential Challenges and Misconceptions

Often systems of equations are determined from contextual situations and so are represented by equations in the standard form. This form is difficult for many students to graph, because they have only graphed equations in slope-intercept form. Students need to have various experiences working with equations that are written in both forms, recognizing when each form is appropriate, and moving between the two forms. They must also have experiences graphing related equations, or linear systems, and understanding what those graphs represent: systems whose graphs are parallel lines have no solutions, systems whose graphs are intersecting lines have one solution, and systems whose graphs are the same line have multiple solutions. Linear systems present many challenges, and students need plenty of guided practice and applications to make sense of them.

## In the Classroom

One teacher begins his lesson on systems of equations by projecting and asking students to solve the following equation puzzle (page A2-21 in the appendix) on the whiteboard.

*Each shape has a different value. Some shapes have the same value in all the equations. What is the value of expression D?*

A. ⬠ + ◇ + ◯ = 26

B. ⬠ + ◯ + ◯ = 23

C. ◇ + ◯ = 15

D. ⬠ + ◇

⬠(11) + ◇(9) + ◯(6) = 26

⬠(11) + ◯(6) + ◯(6) = 23

◇(9) + ◯(6) = 15

⬠(11) + ◇(9) = ? 20

This teacher believes that this equation puzzle will help him understand how his students think about a system of equations prior to a formal lesson. He invites student volunteers to share their thinking and solutions.

Lucy shows her solution and explains, "I looked at the equation C first and saw that the rhombus and the circle equal 15. So in equation A, I subtracted 15 from 26 to get the value for the pentagon, which is 11. Then I looked at the equation B. I figured that the two circles have to equal 12, and the only way to get 12 with doubles is for the circle to be 6. Back to equation C—the circle is 6, so the rhombus has to be 9, because the sum is 15. The pentagon is 11 and the rhombus is 9, so the sum in expression D is 20." This teacher asks if anyone got a different answer or did it differently. Some students say they got stuck and didn't finish the problem.

This teacher then presents these equations and explains that they are related in the same way as the shapes in the puzzle: the values of the variables are different, but each variable has the same value in both equations. He asks his students how they might solve for $f$ and $g$.

$$g = 3f + 6$$

$$2f + 4g = 52$$

He gives his students about two minutes to think of a strategy, and then records their ideas on his conjecture board.

Conjectures

- Make the $f$ and $g$ into $x$ and $y$ then solve.
- Rewrite $2f + 4g = 52$ in $y = mx + b$ form.
- Substitute $3f + 6$ for $g$ in $2f + 4g = 52$ and solve.
- Graph them both.

He is not surprised that his students suggest they rewrite both equations in the same form; however, he is surprised that one suggestion is to change the variables.

He tells students to use a strategy that makes sense to them and sets them to work. As his students work, he notices that the main stumbling block is the different coefficients for $f$ and $g$. He also realizes that his students are not familiar with the differences between the $ax + by = c$ and $y = mx + b$ forms of the equations or how to transition between them. He also observes that the use of $f$ and $g$ rather than the familiar $x$ and $y$ causes many problems. After an allotted time, he invites students to share their work.

### *Student A*

Dougie volunteers to explain the work done by Student A. He states, "First the 2 was factored out from the first equation—no, first the variables were changed from $f$ and $g$ to $x$ and $y$. Then, the 2 was factored out of the first equation so the numbers left with were easy to work with. Now, because $3x + 6$ is the same as $y$, the person substituted $3x + 6$ for y in the first equation. So he got $x + 2(3x + 6) = 26$. He did the distributive property next and got $x + 6x + 6 = 26$. Then all he had to do was solve for $x$ and he got $\frac{20}{7}$." Dougie added, "I think there is a mistake, because I didn't get that answer."

$2F + 4g = 52$
$g = 3F + 6$

$2x + 4y = 52$
$y = 3x + 6$

$x + 2y = 26$
$y = 3x + 6$

$x + 2(3x + 6) = 26$
$x + 6x + 6 = 26$
$7x = 20$
$x = \frac{20}{7}$

### *Student B*

$2F + 4g = 52$
$g = 3f + 6$

$2F + 4(3f + 6) = 52$
$2F + 12F + 24 = 52$
$-24$
$14F = 28$
$F = 2$

Rosie can hardly contain herself as she squirms in her seat waiting to be called on. She insists that she wants to explain her own work and that she is Student B. The first thing she states is, "We should use the variables given to us. I used $f$ and $g$, like in the problem. Then I substituted in for $g$ and did the distributive property and I did it right! I got $12f + 24$. Student A got that part wrong and that is why our answers are different. The correct answer is that $f$ is 2 and $g$ is 12.

### *Student C*

Deion chimes in to support Rosie. He explains, "Student C changed the variable to $x$ and $y$, and so did I because I am more comfortable with them. Student C did everything that Student A did, but, like Rosie, Student C did the distributive property right. When multiplying 2 times the quantity of $x + 6$, Student C got $2x + 12$, and after substituting, Student C got $x = 2$ and $y = 12$."

$$2x + 4y = 52$$
$$y = 3x + 6$$
$$x + 2y = 26$$
$$y = 3x + 6$$
$$x + 2(3x + 6) = 26$$
$$x + 6x + 12 = 26$$
$$7x = 14$$
$$x = 2$$

This teacher makes a habit of posting student work without names and invites other students to explain what happens in the work. Finding and explaining errors involves a higher cognitive demand than just saying something is right or wrong. It also involves most of the eight SMPs.

After discussing the student work, this teacher assigns the *Linear Systems* reproducible on page A2-22 in the appendix.

## Meeting Individual Needs

Many students struggle with linear systems of equations. Because they might solve them by elimination, graphing, or substitution, and because no method lends itself to every problem situation, students need exposure to different situations to help them determine when to use which method. It may be helpful to pose several systems with equations in different forms and just ask the students to identify which method makes the most sense for each system. Then discuss the attributes of the problem situation and why a particular method might be better than another. It may also be helpful to present equations with variables other than $x$ and $y$.

## Additional Reading/Resources

Hedin, David. 2007. "Connecting the Mobiles of Alexander Calder to Linear Equations." *Teaching Mathematics in the Middle School* 12 (8): 452–461.

# 17. Systems

**DOMAIN:** **Expressions and Equations**

**STANDARDS:** **8.EE.8.B.** Solve systems of two linear equations in two variables algebraically, and estimate solutions by graphing the equations. Solve simple cases by inspection.

**8.EE.8c.** Solve real-world and mathematical problems leading to two linear equations in two variables.

## Potential Challenges and Misconceptions

Traditionally, students are taught each method for solving systems of equations in isolation. Often students master the mechanics of each method but are unable to explain how one method relates to the others. Students who are able to see the connectedness among the various methods typically demonstrate a greater understanding of how systems operate. It is crucial that students have multiple opportunities to determine whether to set two slope-intercept forms equal to one another as in some distance, rate, and time problems, or to combine the slope-intercept form with the standard form through substitution, to graph the system, or to use the elimination method. Empowering students to make their own decision and to justify that choice is a crucial step in developing confident mathematics students.

## In the Classroom

Because so few of his students demonstrate proficiency in dealing with equations written in standard form, one teacher begins his class on systems of equations by assigning the *What Is Missing?* reproducible from page A2-23 in the appendix. He wants to confirm that all his students understand how to write an equation given a series of points, find an ordered pair given an equation in standard form with either an $x$- or $y$-value given, and substitute values into a linear or nonlinear equation. He invites his students to check their work by placing large cubes or sticky notes on the floor-sized grid to prove or disprove they have written a linear equation. He does this activity before moving on to simultaneous equations or systems of equations.

This teacher begins his lesson on simultaneous equations by posing the following problem.

> *I have twenty-five dimes and quarters in my pocket. The sum of the coins in my pocket is $5.20. How many dimes do I have? How many quarters do I have?*

This teacher provides plastic coins to his students and gives them a chance to play around with the problem for an allotted time before regrouping to discuss strategies. As his students work, he notices that guess-and-check is the most prevalent strategy. This strategy is followed by organized lists.

When he brings the class together, this teacher asks students to identify their strategy, and he records each on the board. He emphasizes that he is more interested in the strategies they use than their solutions at this time. Just as he observed, his students state that they guessed or made a list.

Donovan adds, "I solved a simpler problem. I made the amount of coins 10 and the value of the coins $1.00 to see how many different ways I could make 100 cents, but then I saw there is only one way so that didn't work."

Tameka says, "I made a table and guessed and checked." The benefit of listing strategies is helpful for those students who do not know what to do or how to begin. Now, they are able to select a strategy recorded on the board.

This teacher suggests his students return to their groups and continue their work on the problem. While the class works on solving the problem, this teacher crafts a graphic organizer for students who need help organizing their data and solving the problem efficiently. When he sees students looking totally bewildered, he shares his graphic organizer with them so they too may have entry into solving the problem successfully. See the reproducible *Graphic Organizer for Simultaneously!* on page A2-24 in the appendix. After an allotted time, this teacher invites student volunteers to share their work and solutions.

### Student A

This student makes an organized list and works with cents. The student appears to divide the twenty-five coins into thirteen dimes and twelve quarters, calculate the value of the dimes, and add that value to the value of the quarters. Because the total cents is less than the desired amount, the student appears to switch the numbers to twelve dimes and thirteen quarters. This gives fifteen cents more to the total but still not enough to equal 520 cents. From there, the student apparently tries different combinations until the sum is 520 cents. The evidence in this student work shows a logical and sequential approach to solving the problem.

| 10¢ | | 25¢ | | 520¢ |
|---|---|---|---|---|
| 13 | | 12 | = | 25 coins |
| 130¢ | + | 300¢ | = | 430¢ |
| 12 | | 13 | = | 25 coins |
| 120¢ | + | 325¢ | = | 445¢ |
| 10 | | 15 | = | 25 coins |
| 100¢ | + | 375¢ | = | 475¢ |
| 8 | | 17 | = | 25 coins |
| 80¢ | + | 425 | = | 505¢ |
| 7 dimes | | 18 quarters | = | 25 coins |
| 70¢ | + | 450 | = | 520¢ |

### Student B

This student solves a simpler problem but appears to have incorrectly determined that five quarters is worth fifty cents. Had the student correctly calculated the five quarters at $1.25, it would be interesting to see what the next step would be. This student would benefit from understanding when it is appropriate to solve a simpler problem and when it makes more sense to work with the data given.

$.10 $.25 $1.00

5(10) 5(25)

.50 + .50 = 1.00

### Student C

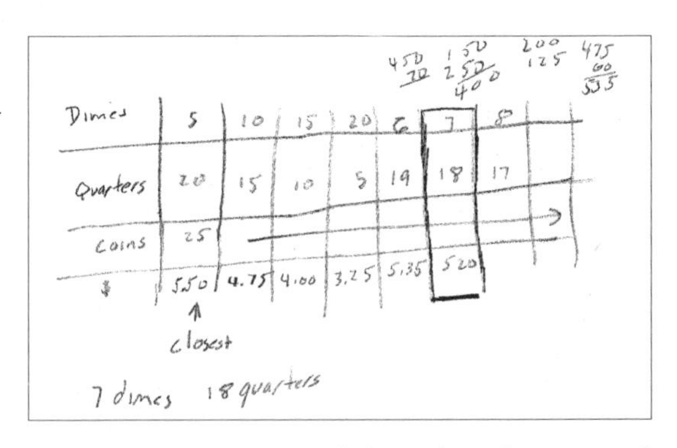


This student makes a table and chooses a variety of combinations of coins that sum to twenty-five. This student begins with five dimes and proceeds to increase the number of dimes and decrease the number of quarters by increments of five. At five quarters, the total value of the coins reaches its lowest total, so it appears that the student goes back to having fewer dimes and more quarters until the $5.20 is reached at seven dimes and eighteen quarters. This student is using an organized, logical, and sequential strategy for solving the problem. This table is similar to the graphic organizer this teacher created. This teacher does not mandate his students use the graphic organizer, but he makes it available for all students and encourages students who are struggling to use it. This teacher also recognizes how challenging it is for many of his students to embrace SMP 1: Make sense of problems and persevere in solving them. He finds it helpful to ask pointed, engaging questions to students who do not appear to be focused to entice them to continue working on the solution.

Next, this teacher assigns the following problem and suggests his students try using the graphic organizer. Because he did not mandate it, some students use it and others choose to solve the system in different ways.

> *At the Hanover Day Celebration, Ami counted the number of bicycles and tricycles lined up for the bicycle parade. She counted 45 handlebars and 111 wheels. How many bicycles are there? How many tricycles are there?*

### Student A

This student makes a table. The evidence indicates that because there were two types of rides, the student divides the number of handlebars by two and then does an organized guess-and-check. This student does not justify the solution. This student chooses to solve the problem without using equations. Although the correct solution is found, the process will not aid in solving simultaneous equations.

| Bike | Trike | Handlebars | Wheels |
|---|---|---|---|
| | | 45 | 111 |
| | | $\frac{45}{2} = 22\frac{1}{2}$ | |
| 24 | 21 | 45 | 48 + 63 = 111 |
| 22 | 23 | 45 | 44 + 69 = 113 |
| 21 | 24 | 45 | 42 + 72 114 |

### Student B

This student makes a copy of the graphic organizer. This student shows evidence of understanding not only how to use the organizer but also how to isolate a variable when working with a system of equations. Notice this student recognizes that if she or he simply subtracts one equation from the other, the resulting equation will still have two variables. Upon recognizing that, the student multiplies one equation by −2 to eliminate the *t* term. This student shows evidence that he or she understands how to represent a situation with related equations, solve simultaneous equations by elimination, and substitute the solution of one equation into the other to find both unknowns.

| | # of | # of wheels | |
|---|---|---|---|
| bikes | b | 2b | (24) |
| Trikes | t | 3t | (21) |
| Total | 45 | 111 | (45) |
| | b+t=45 | 2b+3t=111 | |

$b + t = 45$
$2b + 3t = 111$
$-b - t = -45$
$b + 2t = 66$

$2b + 3t = 111$
$-2b - 2t = -90$
$t = 21$
$\therefore b = 45 - 21$
$b = 24$

$b = 45 - T$
$2b = 111 - 3T$
$b = \frac{111}{2} - \frac{3T}{2}$
$45 - T = \frac{111}{2} - \frac{3T}{2}$
$90 - 2T = 111 - 3T$
$+3T \quad +3T$
$90 + T = 111$
$-90 \quad -90$
$T = 21$
$b = 45 - T$
$b = 45 - 21$
$b = 24$

### Student C

The evidence in this student work illustrates a comfort level with rewriting an equation in slope-intercept form. Once the student isolates the *b* term, this student uses substitution to solve for the *t* term. This student demonstrates an understanding of how to use the substitution method for solving simultaneous equations. The student also shows an ability to work with fractions. This student should be encouraged to define the variables and to check the solutions.

Once this teacher feels certain that most students are able to solve these problem types using more than guess-and-check, he asks Student B to share how he solved the problem with the graphic organizer and the elimination method. He figures that struggling students are more likely to see the value of the graphic organizer when they see a classmate using it to solve the problem accurately and efficiently. After sharing the various solutions to the tricycle and bicycle problem, this teacher assigns the following problem.

*Nik, Mitch, and Matt manufacture tripod (three-legged) and unipod (one-legged) stands for cameras and binoculars. Both types of stands use the same legs and the same "platform" to which the camera or binoculars is attached. They counted twenty-five platforms and fifty-nine legs in the warehouse. How many unipod stands and tripod stands should they make to use up all of the platforms and all of the legs?*

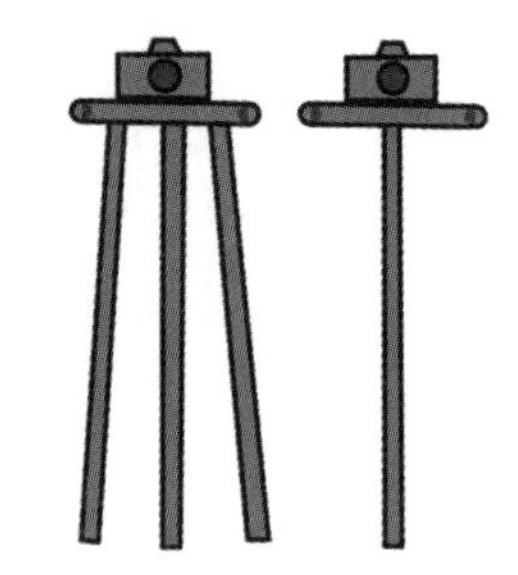

Because this teacher believes his students should do the work first, followed by class sharing and finally his summary, he does not offer assistance while the students work. This "you-we-I" approach empowers his students to believe they are capable of doing the mathematics.

As students take responsibility to struggle with the problems, he notes most are employing several SMPs. This teacher circulates as his students work and poses questions to individual students to keep them engaged, clarify any misunderstandings, or refocus their attention.

After the students have had ample time with this problem, this teacher invites student volunteers to share their work. After the discussion of the solutions, his teacher assigns *Simultaneously!* on page A2-25 in the appendix.

## Meeting Individual Needs

Inviting students to use the floor-sized grid as they work with two points or a point and a slope when writing equations is extremely helpful in letting students visualize how they might find the $x$- and $y$-intercepts and thus write the equations.

Helping students structure the data in mixture word problems is probably the most helpful tool for those students who struggle with organizational skills or who struggle with discerning important information in a problem. The graphic organizer on page A2-24 in the appendix, *Graphic Organizer for Simultaneously!*, does a nice job of helping students who struggle find success in problem solving with simultaneous situations.

## Additional Reading/Resources

Allen, Kasi C. 2013. "Problems Before Procedures: Systems of Equations." *The Mathematics Teacher* 107 (4): 286–291.

# 18. Proportions on the Cartesian Coordinate Plane

**DOMAIN: Expressions and Equations**

**STANDARD: 8.EE.5.** Graph proportional relationships, interpreting the unit rate as the slope of the graph. Compare two different proportional relationships represented in different ways.

## Potential Challenges and Misconceptions

Graphing ratios on the Cartesian coordinate plane is often overlooked as a visual representation that enables students to develop a fundamental understanding of how they might compare slopes, determine which equations are greater than others, and use the graphs to identify shared regions versus regions that are greater than or less than those shared regions. Comparing proportional relationships is often taught through cross multiplication or comparing common denominators. The fact that these methods are taught by rote rather than by helping students understand scale factors and unit rates causes many students great difficulty. Students who are taught to examine double number lines as a representation for proportional relationships, or to graph those relationships on the Cartesian coordinate plane, are exposed to the foundational concept on which slope as ratio is built. In addition, many students struggle when determining how to evaluate inequalities graphed on the Cartesian coordinate plane. Because the slope in an equation is a ratio, it is important that students recognize that they can graph multiple slopes on one set of axes and use the graph to decide how to shade in the appropriate regions.

## In the Classroom

One teacher begins her class by projecting the *Hot Chocolate* reproducible (page A2-26 in the appendix) and asking her students to determine how they might solve the problem. Note that she does not seek an answer at this time.

> *Avery and Connor are making hot chocolate for the cross-country ski team. They are trying to determine which of the following mixes they should use so the hot chocolate is really rich and chocolatey.*

| Parts Cocoa | Parts Milk |
|---|---|
| 3 | 5 |
| 4 | 7 |
| 5 | 8 |
| 7 | 11 |

> *What strategies might you use to determine which of the mixtures Avery and Connor should use?*

SECTION 2: EQUATIONS

After a few minutes, this teacher begins to record the various strategies her students suggest on the board. Donovan replies, "I would find the common denominator and make equivalent fractions." McKenzie says, "Gee, I think I'd find the difference between the parts milk and parts cocoa." Emma adds, "Hmm, I would change the ratios to decimals." This teacher makes note that McKenzie is thinking additively instead of multiplicatively. She will address that misconception at some point, but for now she wants to extend her students' thinking to include the ideas of using common numerators and graphing the relationship between the cocoa and the milk.

Once students' strategies have been recorded, this teacher distributes graph paper and tells her students to graph the ratios. She purposely does not give them any more information than that. She does not tell them to graph the part-to-part relationship or the part-to-whole relationship but, rather, wants to see how her students are thinking about the situation. After an allotted time, she invites student volunteers to the document camera to share their graphs.

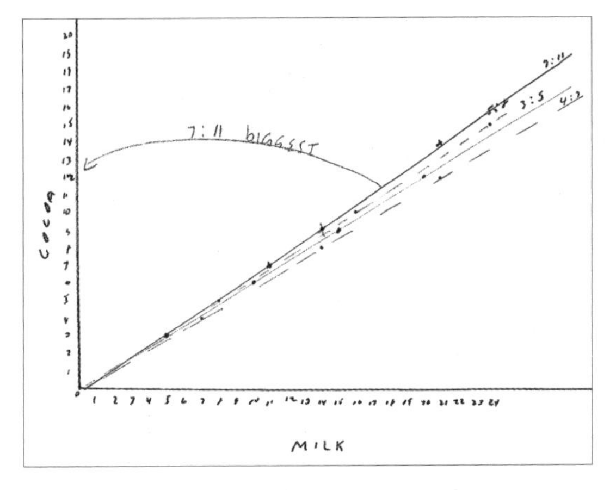


Abby begins by projecting hers. She explains, "I did the cocoa to milk. I can see that the mixtures are really close to one another, but the 4:7 mixture is the milkiest. The 7:11 ratio is the most chocolatey. I know that 'cuz the line is the closest to the cocoa axis."

McKenzie raises her hand and states, "Gee, I thought that the 3:5 ratio would be the best because there are only 2 spaces between 3 and 5. There are 3 spaces between 4 and 7 and 5 and 8, and there are 4 spaces between 7 and 11. I thought the closer the numbers the bigger the value." At this time, this teacher suggests that McKenzie make a rate table for each of the mixtures to see how they "grow" multiplicatively. She suggests that for each new equivalent ratio, McKenzie indicate the scale factor that she is using to go from, say, 7:11 to 49:77.

Noah shares his graph and explains, "Mine looks different from Abby's. I made the milk on the $y$-axis, so my steepest line is for the milkiest mixture. I agree with Abby that the 7:11 is the most chocolatey mix."

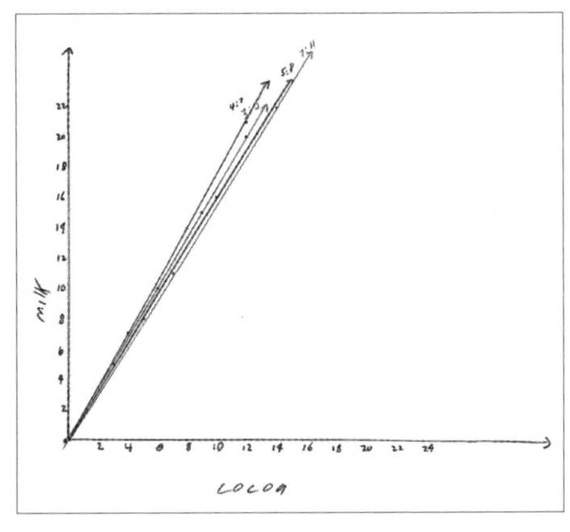


Once the students are afforded the opportunity to share any differences, this teacher moves on to asking her students to find two ratios that are greater than $\frac{2}{3}$ but less than $\frac{3}{4}$, using any strategy. Kenny almost jumps out of his seat to share his work. He explains, "First I tried to make a table, but I got confused so I moved to common denominators. That worked, and I can list a whole bunch of ratios that come between $\frac{80}{120}$ and $\frac{90}{120}$. But then I thought about graphing the two ratios, and look at this. You can see lots of ratios that fit on the points between $\frac{2}{3}$ and $\frac{3}{4}$. Just by looking you can tell the answers."

After Kenny finishes explaining his graph, this teacher asks whether anyone did it differently or got a different answer. When no one responds, she preassesses her students' understanding of distance, rate, and time situations before assigning the next problem. (Depending on the length of your class period, this may be done the next day. For 90-minute classes, this is done on the same day.)

This teacher presents this problem and asks students to turn to a shoulder partner to discuss ideas.

> *Describe how you might determine how fast, on average, a vehicle might be traveling if you know the distance traveled and the time it took to cover that distance.*

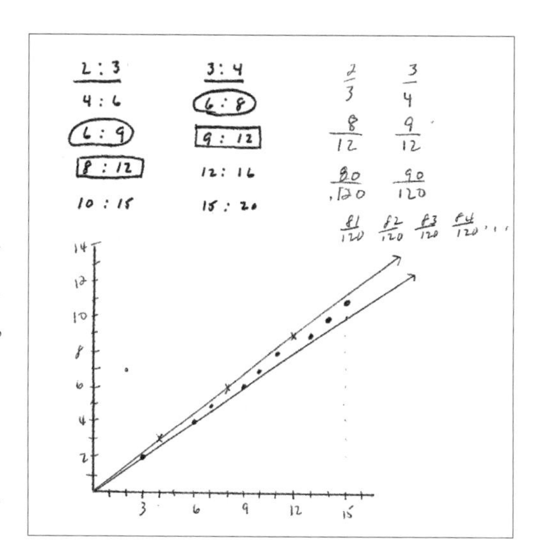


After an allotted time, she invites student volunteers to share their thinking. She records their suggestions on the board. They include: "Look at the speedometer in your car." "You need to think about the fact that the speed is related to the distance and time, so you either multiply or divide the distance and time to get the speed. I forget which." "I remember that distance equals rate times time, so to find rate you divide the distance by the time." "Write an equation and plug in the numbers."

Once all suggestions have been recorded, she assigns triads of students to solve the following problem (also found on the *Stephen and Cole* reproducible on page A2-27 in the appendix):

> *Stephen drove at a constant speed from town X to town Y at 9:00 a.m. yesterday. Half an hour later, Cole drove from town X to town Y at a constant speed that was 30 km/h faster than Stephen's. By 9:30 a.m., Stephen had already traveled 40 kilometers. Cole caught up with Stephen at town Y, arriving about the same time as Stephen.*
>
> *a. At what speed was Stephen driving?*
>
> *b. What was the distance between the two towns?*

This teacher observes each group as they work, listening intentionally for the strategies the students try, the language they use, and how well they listen to each other and support every member of the group. She notices that Evan's group began with a graph. Julia suggests, "I think the axes should be labeled *distance* and *time*, and the distance depends on the time." Evan adds, "That makes sense but how should we label the time? By the time on the clock or by hours? I guess we should start at 9:00, 'cuz by 9:30 Stephen was already 40 kilometers away from where he started. At 10:00 Stephen will be at 80 kilometers and Cole would be at 55 kilometers."

At Carter's group, Mona is heard saying, "If we use the distance equals rate times time equation, we can find that Stephen traveled at an average of 80 kilometers per hour, so Cole went at 110 kilometers per hour. How can we find when they meet up?"

Carter suggests, "We also know that Cole started half an hour later, so at 9:30. Let's see if we can graph that."

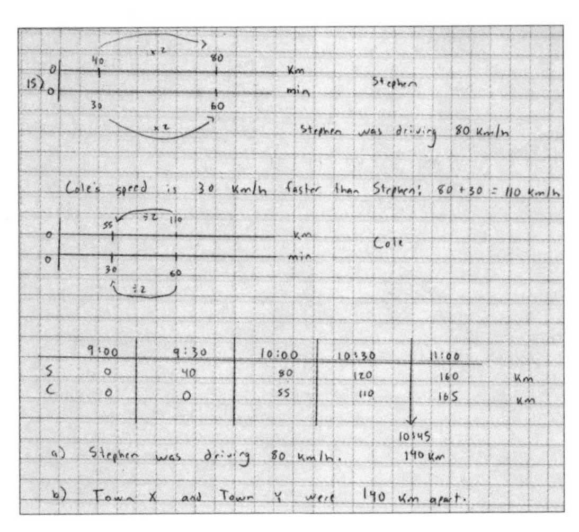


Mona adds, "I don't know how a graph would look. We could make a table though and see when they travel the same distance."

After an allotted time, this teacher invites volunteers to share their work. Mona asks to go first and shares her table and double number lines.

Mona explains her double number line first. She states, "I know that Stephen took 30 minutes to drive 40 kilometers, so I marked that ratio on the double number line. Next, I needed to find how far he went in 1 hour, so I multiplied both values by 2 and got 80 kilometers in 60 minutes. Because Cole was driving 30 kilometers an hour faster, I added that to the 80 kilometers to find that Cole was driving 110 kilometers per hour. My table shows the distance each person traveled every $\frac{1}{2}$ hour. I struggled to figure out what the distance between the towns was until my group suggested we look at distances every 15 minutes. When we did that, I found that the towns were about 140 kilometers apart."

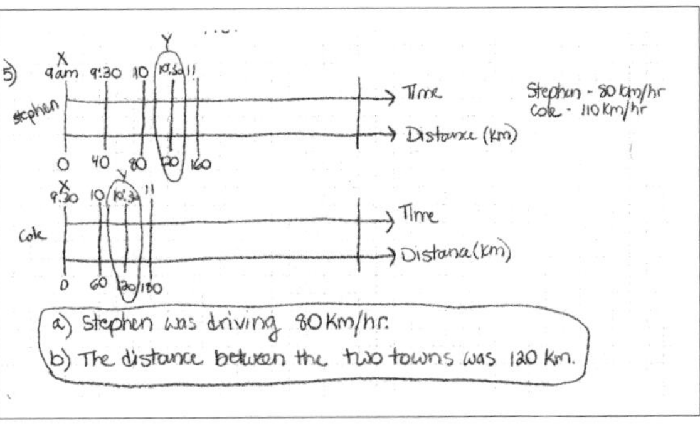


Lebron's group also makes a double number line and correctly calculates the rate at which both Stephen and Cole were driving. However, they make a calculation error when finding the distance. If Cole were traveling at 110 km/h, then in a half hour he travels 55 kilometers, not 60. This group shows conceptual understanding but needs to be reminded to check their calculations.

Evan's group began by making a graph. He explains, "Our group graphed Stephen's distance every hour starting at (0, 0). Then we graphed Cole's distance starting at half an hour. We saw that they intersect at about 140 kilometers in just less than 2 hours."

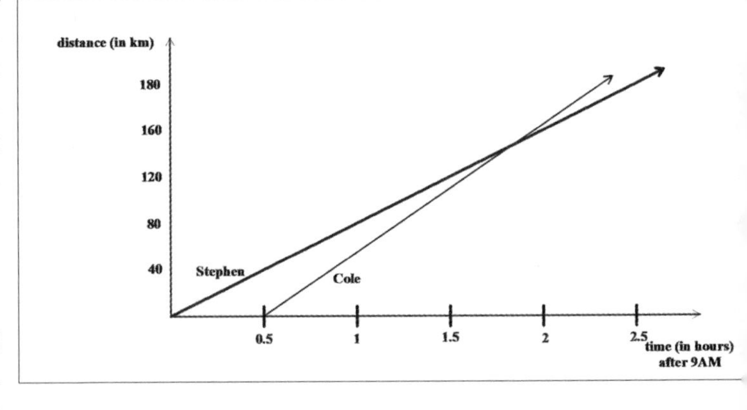


## Meeting Individual Needs

Many students will benefit from using the graph to compare ratios. This is a good representation to help students understand that the graph actually represents a ratio between two variables. Similarly, the double number line supports the graph and the table. Some students may choose the representation that makes the most sense to them and to use that representation consistently. The double number line is a visual explanation of why cross-multiplication works and should be encouraged as the go-to strategy, rather than having students cross-multiply alone.

## Additional Reading/Resources

Ozgun-Koca, S. Asli, Thomas G. Edwards, and Kenneth R. Chelst. 2013. "Mathematical Explorations: Exercise Away the Big Mac: Ratios, Rates, and Proportions in Context." *Teaching Mathematics in the Middle School* 10 (3): 184–186.

# 19. What Is the Better Buy?

**DOMAIN: Expressions and Equations**

**STANDARDS: 8.EE.8b.** Solve systems of two linear equations in two variables algebraically, and estimate solutions by graphing the equations. Solve simple cases by inspection.

**8.EE.8c.** Solve real-world and mathematical problems leading to two linear equations in two variables.

## Potential Challenges and Misconceptions

Comparing costs, shipping, and value between and among many options is a crucial component in our society. Many students fail to recognize that having a solid understanding of how to compare equations, graphs, and tables is an important skill that will affect them in their daily lives. Students need multiple and varied examples and situations that represent data in a variety of forms. With this experience, they'll build the confidence to sort through the data and data representations they will face when making major purchases or finding the best deal.

SECTION 2: EQUATIONS

## In the Classroom

One teacher begins by projecting the Pi Day Celebration problem on the board, found on the *Which Region Is It?* reproducible on page A2-28 in the appendix.

> *Sophia is in charge of ordering pizzas and cinnamon breadsticks for the Math Club Pi Day Celebration. A local pizzeria is offering a special deal for party orders: If you order a total of at least thirty-two items, the price will be $10 per pizza and $4 per order of cinnamon breadsticks. Sophia has a budget of $200. How many pizzas and breadsticks might Sophia order?*

This teacher challenges her students to solve the problem. She is interested to know whether or not her students will get one solution or offer many possibilities.

As she walks around the room, she records on an observation protocol which students are approaching the problem by making an organized list, writing equations or inequalities, graphing the data, or making a table. She uses an observation protocol (see a sample on page A2-29 in the appendix) to identify individual student strengths and weaknesses and to decide her next instructional steps. When she sees some students simply writing disorganized computations, she joins those students and asks them to explain to her what they are thinking. She does not accept the most common response, "I don't know," but prods the students to read the problem aloud and to tell her what the problem is asking and how they might represent the data to answer the question posed. She has discovered that when she asks her students to articulate their thoughts, they often clarify for themselves what the problem is asking and how they might go about solving it.

This teacher invites Josh's group to share their work. Josh explains, "We made a table and used guess-and-check to see how many pizzas and breadsticks we could buy. We started with no pizzas and thirty-two breadsticks, and that cost $128. Then we tried seven pizzas and thirty-two breadsticks, and that cost \$198. Finally, we tried sixteen pizzas and ten breadsticks and that cost exactly \$200 but was only twenty-six items, so that didn't work."

Chonda volunteers to share her group's work. "We made up two equations, $p + b = 32$ and $10p + 4b = 200$. We solved for $b$ and got 12, so that meant that $p = 20$. But the words *at least* in the problem means we need to use the greater than or equal to symbols for the amount and less than or equal to for the money. We think it means that $p + b \geq 32$ and $10p + 4b \leq 200$."

Simon's group agrees with Chonda's group on writing the equation, but he says, "We got $b \geq 20$ and $p \leq 12$. I think that Chonda's group made a mistake somewhere, because twenty pizzas would cost \$200 and there would be no money left for breadsticks."

After discussing the students' solutions, this teacher then projects the following inequalities and asks her students to determine which best represents the data in the problem. She challenges them to create a mathematical argument for the choice they make and to prove the other inequalities do not meet the criteria in the problem.

| | | | |
|---|---|---|---|
| $10p + 4b \leq 200$ | $10p + 4b \leq 32$ | $10p + 4b \geq 200$ | $10p + 4b \geq 32$ |
| $p + b \geq 32$ | $p + b \geq 200$ | $p + b \leq 32$ | $p + b \leq 200$ |

After students share their solutions, this teacher further challenges them to graph the inequalities. For many students, this is the first time they have been asked to graph inequalities on the Cartesian coordinate plane, so this teacher distributes large easel paper for each small group. Once the graphs are completed, she instructs students to hang the graphs around the room for a facilitated gallery walk. Not surprising is the inclusion of dotted lines, solid lines, and shading. There are also graphs without the necessary shading.

These differences help this teacher engender a rich conversation about the mathematical significance in the graphs. She asks her students what they believe the difference is between the dotted and solid lines. Avery offers, "I think the dotted lines are like the open circles on the number line. They mean the values come close to but are not equal to where the lines are."

Spenser agrees and adds, "Remember, when we graphed ratios, all the equivalent ratios were on the same line, so when the line is solid it means the solution values equal those ratios."

This teacher also asks her students how they know what regions they should shade. Danicka responds, "I have trouble sometimes deciding when I put both graphs on the paper at one time. I think I get them right when I graph one at a time and shade in just that one area. Then, if I shade in the second line and shade in the area I just look for where the shaded parts overlap. I think that is best for me."

Marshall adds, "I need to use colored pencils to see the parts that overlap. It's too hard to see when all I use is pencil."

After the discussion, this teacher assigns the remaining problems on *Which Region Is It?* and the *Solving Situations* reproducible from A2-30 in the appendix.

## Meeting Individual Needs

For students who struggle graphing the equations, it is helpful to begin by having them graph two ratios, such as three parts orange to five parts water and four parts orange to seven parts water. Once those ratios are graphed, instruct them to shade in the area with ratios that are greater than 4:7 but less than 3:5. The shaded area will be between the two lines. Next, ask them to identify all the ratios greater than 3:4 or less than 4:7. This practice will help students understand what the shaded areas represent. After working with simple ratios, progress to more contextual problems.

## Additional Reading/Resources

Collins, Anne, and Linda Dacey. 2011. *Zeroing in on Number and Operations, Grades 7–8*. Portland, ME: Stenhouse.

# 20. Walking Rates

**DOMAIN: Expressions and Equations**

**STANDARDS:** **6.RP.3.** Use ratio and rate reasoning to solve real-world and mathematical problems, e.g., by reasoning about tables of equivalent ratios, tape diagrams, double number line diagrams, or equations.
**7.EE.4.** Use variables to represent quantities in a real-world or mathematical problem, and construct simple equations and inequalities to solve problems by reasoning about the quantities.

## Potential Challenges and Misconceptions

Too often, much of the algebra in which students are involved consists of symbolic representations or uninteresting problems that have little to no appeal to students, even though students who have opportunities to simulate actions are more likely to translate those experiences to similar situations. Because many students taking algebra are close in age to getting their driver's permit or license, problems that relate to speed may be interesting. Putting distance, rate, and time equations into interesting scenarios usually captures students' interest and their desire to solve the problems as accurately as possible. Students often memorize the formula distance equals rate times time but are unable to apply it appropriately in various situations. These are just some examples of why it is necessary for students to understand what the equations represent and how to work flexibly with them.

## In the Classroom

One teacher begins this investigation into distance, rate, and time by laying down five meter sticks in a row at five different stations around the room. She provides each group of four students with a timer and a recording sheet. She tells the students that they will be investigating how fast they walk, or their individual walking rates. Students are told to take turns being the walker and the timer. This teacher demonstrates how the students will conduct the experiment by asking a student volunteer to stand at the beginning of the row of meter sticks. She serves as the timer and when she says, "Go," the student walks to the end of the meter sticks and says, "Stop." This teacher announces the time it took, and all students in the group record it in a table. When groups understand the process for timing and recording, she asks students to get a walking rate for each student. She challenges groups to determine which student walks at the fastest rate and to write an equation for the distance, rate, and time. She also tells the students to graph each student's results.

Upon completion of this module, this teacher assigns this problem (also found on the *Backpacking* reproducible on page A2-32 in the appendix):

*Jordan went for a hike on Mt. Tom. When he got to the top of a cliff that is 121 meters high, he accidentally dropped his backpack off the edge. His friend Nicole was at the bottom of the cliff 43 meters from the base. She saw what was happening and ran to try and catch the backpack. She ran toward the base of the cliff as fast as she could at the exact moment the backpack was dropped. After 1 second, she was 35 meters away from the base.*

- *Draw a picture to illustrate what is happening.*
- *Model Nicole's distance from the base of the cliff as a linear equation of the time she began running. Define each variable.*
- *Use your model to predict how long Nicole took to reach the base of the cliff. Show your work.*

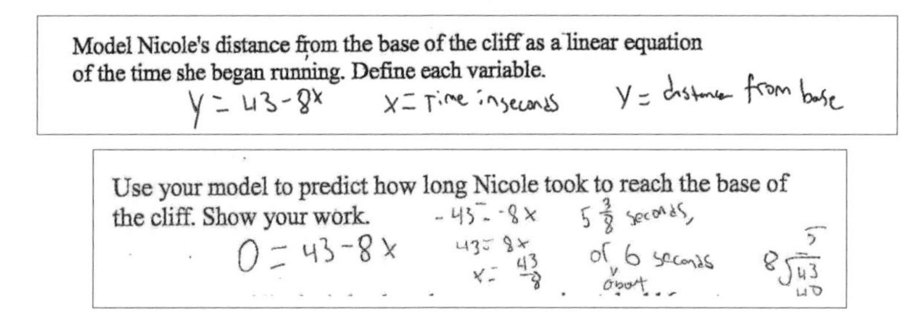
Model Nicole's distance from the base of the cliff as a linear equation of the time she began running. Define each variable.

y = 43 - 8x    x = Time in seconds    y = distance from base

Use your model to predict how long Nicole took to reach the base of the cliff. Show your work.

0 = 43 - 8x    -43 = -8x    43 = 8x    x = 43/8    5 3/8 seconds, or 6 seconds about...

### *Student A*

This student correctly writes the equation for the distance; however, notice that the student unnecessarily converts the variables back to the more comfortable $x$ and $y$ rather than working with $d$, $r$, and $t$. This student correctly determines that it will take Nicole $5\frac{3}{8}$ seconds to reach the base of the cliff.

### *Student B*

This student also correctly calculates the time it will take for Nicole to reach the base of the cliff. This student maintains the integrity of the variables that define distance and rate but chooses $s$ for time in seconds. He or she defines the variables.

Model Nicole's distance from the base of the cliff as a linear equation of the time she began running. Define each variable.

d(s) = 43 - 8s    d = distance    Speed = 8m/sec    S = time nicole ran (seconds)

Use your model to predict how long Nicole took to reach the base of the cliff. Show your work.    = 43 - 8S

$\frac{-43}{-8} = \frac{-8S}{-8}$    S = 5 3/8

This teacher asks whether anyone used a double number line to represent the situation. Although hesitant to share his work, Brett volunteers to share his diagram. He explains, "When I first made the double number line I thought I needed to put both the 35 meters and 43 meters on the distance line, but then I realized that didn't make sense. I needed the unit rate of how far she traveled in one second. So I changed the 35 to the difference between 43 and 35, and because I was moving backward on the number line, I made the 8 negative in my mind. Then I found the unit rate is 8 meters per second. I divided 43 by 8 and got $5\frac{3}{8}$ seconds. I really like seeing what is happening; that's why I did the double number line."

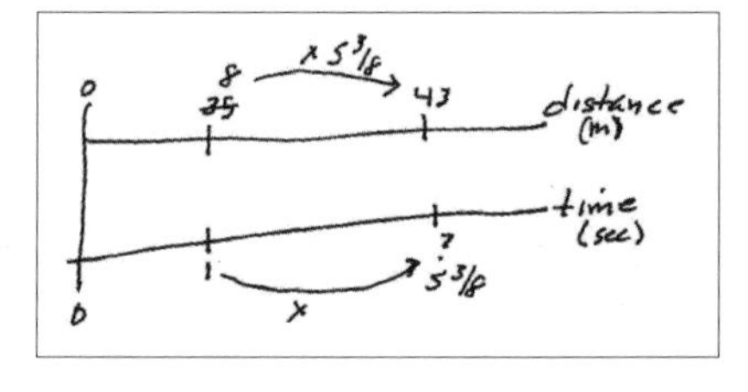


This teacher then assigns the reproducible *How Fast Did I Go?*, page A2-33 in the appendix.

## Meeting Individual Needs

For students who struggle to understand the proportional relationship between distance, rate, and time, the double number line is a valuable tool, because it makes visual the importance of the unit rate and how that unit rate can lead to the relationship between distance and time. Many students will benefit from experimenting with the distance they can walk or run in a given time. This can be done in the schoolyard or on a track if one is nearby.

## Additional Reading/Resources

Kastberg, Signe E., Beatriz S. D'Ambrosio, Kathleen Lynch-Davis, Alexia Mintos, and Kathryn Krawczyk. 2013/2014. "CCSSM Challenge: Graphing Ratio and Proportion." *Teaching Mathematics in the Middle School* 19 (5): 294–300.

# 21. Absolute Value

**DOMAIN:** **Number Systems, Algebra**

**STANDARDS:** **6.NS.7c.** Understand the absolute value of a rational number as its distance from 0 on the number line; interpret absolute value as magnitude for a positive or negative quantity in a real-world situation.
**HSA.SS.B.3.** Choose and produce an equivalent form of an expression to reveal and explain properties of the quantity represented by the expression.

## Potential Challenges and Misconceptions

Imagine students' confusion when, after coming to understand integers, especially negative integers, and focusing on the sign of a product or difference, they are told that the answer is always positive unless, of course, the answer is zero, which is neither positive nor negative. The introduction to absolute value can cause confusion and angst—or both—for teachers and students alike. Add absolute value inequalities, and the scenario is ripe for misunderstandings, partial understandings, and misconceptions. Most students understand the concept that distance is always nonnegative, so absolute value equalities are more readily understandable than absolute value inequalities.

## In the Classroom

Physically modeling absolute value is crucial in helping students begin their journey with absolute value equations and inequalities. One teacher begins his lesson by telling the class to line up in a horizontal row in the middle of the mathematics corridor (after the hallway is cleared of students changing classes!). This teacher tells his students to take four steps. Most of the class moves four steps forward. This teacher tells them to return to their original location and this time to be creative and move four steps. Some move to the right, others to the left. Some move forward, others move backward, some move diagonally. This teacher then asks the students why they did not end up at the same location. Most students think this is a trick question, but they do answer it seriously by stating, "We all started at a different spot, so we all end up at a different spot." Yolanda adds, "And we all moved the same distance, but in different directions." Identifying the starting location is important when working with absolute value, so this teacher is pleased with the response. The class returns to the classroom, and this teacher asks his students to make conjectures about the term *absolute value* and to think about it in terms of the activity they just completed.

Conjecture Board

- *Absolute value is about distance.*
- *Absolute value does not tell what direction we moved.*
- *The starting point varies, so zero is not important.*
- *When we moved left or back we went in a negative direction, but we moved a positive distance.*
- *We need to know the starting point.*

After recording the conjectures, this teacher asks whether anyone can find a counterexample, or negation, for any of the conjectures. Gabby offers, "I want to revise 'The starting point varies, so zero is not important' and make it read, 'The starting point becomes zero!' Ava suggests that they add that absolute value is always positive. This teacher records students' revisions and additional statements on the conjecture board.

Conjecture Board

- *Absolute value is about distance.*
- *Absolute value does not tell what direction we moved.*
- ~~*The starting point varies, so zero is not important.*~~ *The starting point becomes zero!*
- *When we moved left or back we went in a negative direction, but we moved a positive distance.*
- ~~*We need to know the starting point.*~~ *The starting point is important to know.*
- *Absolute value is always positive.*

These conjectures and revisions illustrate that most of the students are beginning to understand the concept of absolute value, but the challenge to work with contexts and the mathematical notation must still be addressed. To do this, this teacher presents the following situation (also found on the *Amanda's Subway Ride* reproducible on page A2-34 in the appendix):

> *Amanda is taking the subway to meet some friends. She asks a fellow passenger how many stops there are until she gets to Park Street Station. The passenger tells her it is seven stops but neglects to tell her in which direction.*

This teacher challenges his students to write an equation and represent it on a number line to show the choices Amanda has to make based on the information the passenger gave her. He asks students to discuss their equations with their shoulder partners. Then he asks Grayson to share his equations and number line.

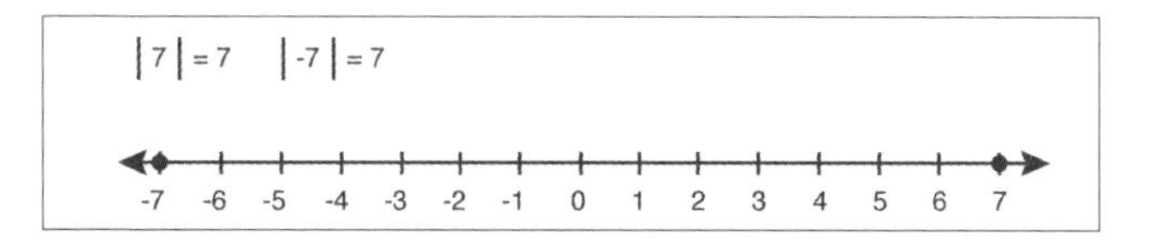


This teacher then asks whether anyone got a different answer or did it differently. When no one chooses to share a different equation, this teacher moves on to more complex equations. He projects the equations $|3b + 2| = 8$ and $|3b + 2| = -8$ and challenges his students to solve the equations and graph each on separate number lines. He provides time for his students to work together, and he walks around to various groups observing how students are thinking about the situation. After an allotted time, this teacher asks volunteers to share their solutions. Maya shares her work.

Maya explains, "I tried to solve the absolute value equation by using the symbols for absolute value, but then I remembered I had to drop them and make two equations, because you can have a positive or negative direction from zero. So what's inside the absolute value symbol, $3b + 2$, can equal $+8$ or $-8$. I got $b = 2$ or $b = -\frac{10}{3}$. I didn't know how to do the second equation."

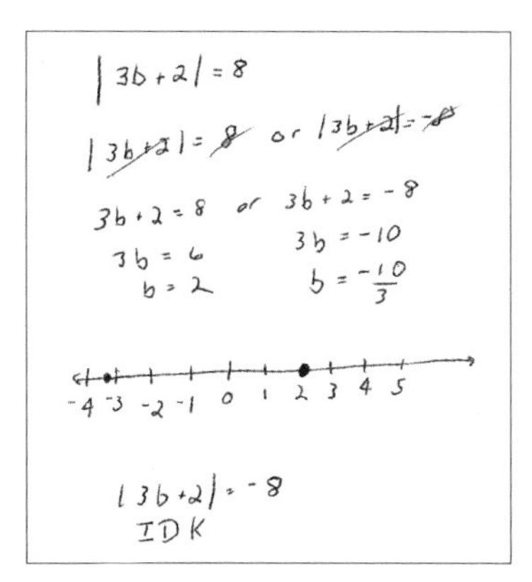

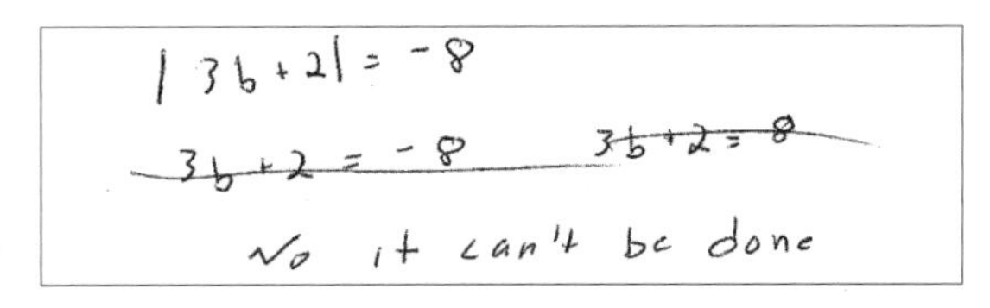


Jared shares his work for the second equation. "I started to solve this one, but then I remembered when we were modeling absolute value in the corridor. The distance couldn't be negative, so this equation has no solutions." There is a chorus of "Aha" and "I get it" from his peers.

Seeing that his students are still hesitant about solving the absolute value equations, this teacher pairs his students and assigns the *My Number Line Representation* cards from page A2-35 in the appendix. (Note that the teacher must cut out the cards on the reproducible and mix them up before giving them to student pairs.) As students work to match the equation to the correct number line, this teacher walks around, stopping to ask students to explain what they are thinking. Not surprisingly, many students need assistance in clarifying the "and/or" representations for the equations and some with the directions of the lines.

## Meeting Individual Needs

When working through the problems posed in class, it is often helpful to provide students with a large number line (desk-sized) and sticky notes and/or cardboard cutouts (e.g., people or cars) that students can actually manipulate on the number line. In the case of going seven stops, one cutout of a person might be placed 7 units to the right of 0 and a second one 7 units to the left of 0. Many students benefit from this type of modeling, because it makes the conceptual more concrete and understandable.

## Additional Reading/Resources

Ellis, Mark W., and Janet L. Bryson. 2011. "A Conceptual Approach to Absolute Value Equations and Inequalities." *Mathematics Teacher* 104 (8): 592–598.

Ponce, Gregorio. 2008. "Using, Seeing, Feeling, and Doing Absolute Value for Deeper Understanding." *Teaching Mathematics in the Middle School* 14 (4): 234–240.

# 22. Nonlinear Explorations

**DOMAIN: Algebra**

**STANDARD: A-SSE.3.** Choose and produce an equivalent form of an expression to reveal and explain properties of the quantity represented by the expression.

## Potential Challenges and Misconceptions

It is challenging for many students to visualize what makes a written equation linear or nonlinear. Too often they lack experience in "playing" with situations, tables, and graphs that would help them discover what makes equations and/or situations linear or nonlinear. A powerful solution is to engage students with inexpensive models that allow them to play with the model, collect and order the data, graph that data, and then determine whether the situation is linear or nonlinear.

## In the Classroom

Because most of this teacher's students have been working extensively with linear equations, he decides to have his students experiment with and model quadratic and other nonlinear equations. He groups his students into triads and gives each group a 2- or 3-liter plastic bottle, a timer, and duct tape. If the weather is inclement or too cold, he also gives each group a large bucket and towel. If the weather is pleasant, he takes his students outdoors and does not have to worry about them making a mess! Students are instructed to fill the bottle with water and to release the water in 10-second intervals. (The larger the bottle, the more distinct the graph will be.) After each 10 seconds, a second student measures the height of the water that is left, and the process is repeated until no more water can be released. (Some water will be left in the bottle due to the position of the hole.) Students record their data in tables. (Full teacher instructions for this activity, *Bottle Experiment Directions*, can be found on page A2-36 in the appendix.)

Once the experiment is complete, students are told to graph their data and chat among themselves to describe their results. As they work, this teacher observes each group, offering suggestions and asking questions. He asks one triad to describe what the $x$- and $y$-intercepts represent. These students recognize that the $y$-intercept represents the starting height of the water in the bottle and that there is no $x$-intercept because there is still water in the bottle, below the hole they punched. At another group, he focuses on the table and asks students to describe how the data are developing—what patterns or relationships they are seeing as they conduct the experiment. At yet another group, he asks the students how they might go about writing an equation.

After an allotted time this teacher calls on various groups to share their work. Avery's group shares their table. She explains, "We did the experiment twice so we could discover what factors would affect the results. We saw that the person covering the hole had to stay the same for the whole experiment. We also saw that the force of the water coming out of the hole got less at each time interval. We conjectured that the equation is not going to be linear, because we don't have a constant slope. So we graphed the data from our second trial and saw that it is not a straight line. The line curves but not like an exponential curve. We are not sure how to write an equation, but we do know the constant term will be the height of the water."

Maddy reports out for her group. She states, "We agree with Avery that this is not linear. We used 5-second intervals for the time. We also made a spreadsheet with our data, and we tested it three times to be sure we had the right data. We also took turns opening the cap and covering the hole but that changed the data. Josh was the best at opening the cap and Lenny was best at covering the hole. We graphed the data and made a slight curve. We need help writing the equation."

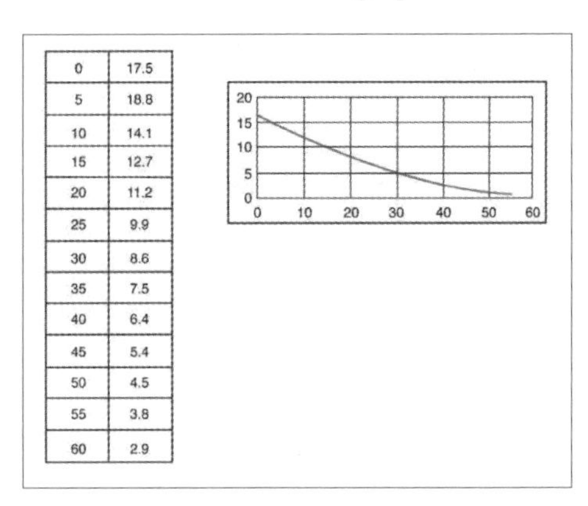

| 0 | 17.5 |
|---|---|
| 5 | 18.8 |
| 10 | 14.1 |
| 15 | 12.7 |
| 20 | 11.2 |
| 25 | 9.9 |
| 30 | 8.6 |
| 35 | 7.5 |
| 40 | 6.4 |
| 45 | 5.4 |
| 50 | 4.5 |
| 55 | 3.8 |
| 60 | 2.9 |



After the experiment and classroom discussion, this teacher shows the students how to use the graphing calculator to draw the graph and write the equation. Each group of students is provided an opportunity to comment on their graphs and their equations.

Following this, he assigns the reproducible *Am I Linear, Quadratic, Exponential, or Other?*, page A2-37 in the appendix, which presents linear and nonlinear tables. He asks students to graph each input-output table and then to find the difference between the first and second outputs, the second and third outputs, and so on throughout the table. If the differences are not constant, he asks students to find the differences between differences and to repeat the process until the differences are constant. He hopes his students will come to realize that when the first difference is constant, there is a linear relationship; when the second difference is constant, there is a quadratic relationship; and when the third finite difference is constant, there is a cubic relationship. This teacher realizes it is important for his students to graph these input-output tables to see for themselves, because presenting the rule without such visual support in the past has not been effective; his students were unable to describe what the graph look like, based on the data in the table. (See the lesson "Why Finite Differences Works" for more about finite difference methods.)

Once this teacher believes his students understand some of the differences between linear and nonlinear relationships, he assigns *Delilah and the Rooftop* from page A2-38 in the appendix. Before the students begin working on this, this teacher explains the inclusion of the force of gravity in the quadratic equation as the $a$ term. He reminds his students that the unit of measure affects the value used to account for gravity. When the measurement is in feet, the value is $-16$, and when the measurement is in meters, the value is $-4.9$. His students were already comfortable with defining and determining the $c$ term.

## Meeting Individual Needs

Students who enjoy a challenge might benefit from the extended problem on the *Will She Catch It?* and *Modeling Motion* reproducibles, pages A2-39 to A2-40 in the appendix.

Because many experiments involving quadratics are graphed in the first quadrant, they do not show the complete parabola. This can be problematic for many students, so it is important that students are able to play with situations that include the first and second quadrants or the third and fourth quadrants. Or at the very least, teachers should extend the domain to include values greater than and less than zero and encourage the students to examine the context to determine which values make sense.

## Additional Reading/Resources

Dahl, Terri, Jerry Johnson, Monique Morton, and Sharon Walen. 2005. "Chapter 2: Algebra." *Mathematics Assessment Sampler: Items Aligned with NCTM's Principles and Standards Grades 9–12*, ed. Betty Travis, series ed. Anne M. Collins. Reston, VA: NCTM.

# Functions

## 23. Functionally Speaking

**DOMAIN:** **Functions**

**STANDARDS:** **F-IF.1.** Understand that a function from one set (called the domain) to another set (called the range) assigns to each element of the domain exactly one element of the range. If $f$ is a function and $x$ is an element of its domain, then $f(x)$ denotes the output of $f$ corresponding to the input $x$. The graph of $f$ is the graph of the equation $y = f(x)$.

**F-IF.2.** Use function notation, evaluate functions for inputs in their domains, and interpret statements that use function notation in terms of a context.

### Potential Challenges and Misconceptions

Students are often told that a function is a relationship in which every $x$-value has one and only one $y$-value. Unfortunately, being given a definition in isolation makes it difficult to retrieve because the definition is not connected to an experience. But when a term is modeled, spoken about repeatedly, and derived from a context, mental connections are made. The term and its definition are kept in a person's long-term memory after they are recalled more than once and thus are more easily retrieved when needed. Students who are told only the definition of *function* often misremember that definition and state that a function is a relationship in which every $y$-value has one $x$-value. For students to avoid the most common misconceptions, it is important for them to hear the term spoken often and appropriately, to bring and share past experiences with others, to see the notation $f(x)$ and hear it read as "eff of ex," and finally to work occasionally with decontextualized functions, or functions for functions' sake. Working with functions for functions' sake can allow students the opportunity to utilize the algebraic properties of the functions without context getting in the way. It is essential that teachers take and provide opportunities to observe and confront misconceptions that arise around function notation.

### In the Classroom

One teacher begins his class by posing the following range question: *What is a function in mathematics and how is it represented?* Recall that a range question is designed to let the teacher understand what students know about a particular concept, what they misremember, or what misconceptions they are bringing to the concept. This teacher records the students' responses and leaves them posted throughout the study of functions. After reinforcing and expanding student understanding, he will ask the students to revise or delete any statements that are not always mathematically true.

This teacher projects the first function from *Working with Functions, Part 1* on page A3-2 in the appendix, $f(x) = 3x - 4$. He asks his students to make a list of what they see. After an allotted time, he asks student volunteers to share their thinking. He records, *an equation*; *a linear equation*; *a slope of 3 and a y-intercept of –4*. No students mention the function notation. This teacher makes a positive comment about what the students offer and then formally introduces the term *function* and the function notation. After mentioning that in a function the domain is the set of inputs and the range is the set of outputs, he tells them that the notation $f(x) = 3x - 4$ means given a number $x$ in the domain, the corresponding output is four less than three times $x$. He asks his students to find the outputs of $f(x) = 3x - 4$ if $x = \{1, 2, 3, 4, 5\}$.

When students had enough time, he asks student volunteers to share their thinking. When a number of students shared the answers $f(1) = -1$, $f(2) = 2$, $f(3) = 5$, $f(4) = 8$, and $f(5) = 11$, Andy says, "I got different answers." He goes to the board and writes:

$$f(1) = 3x - 4(1) = 3x - 4$$
$$f(2) = 3x - 4(2) = 3x - 8$$
$$f(3) = 3x - 4(3) = 3x - 12$$
$$f(4) = 3x - 4(4) = 3x - 16$$
$$f(5) = 3x - 4(5) = 3x - 20$$

Sarah says, "I got something different" and writes $f(2) = (3x - 4)(2) = 6x - 8$ on the board. Abby raises her hand and replies, "But that's not how functions work. If $f$ of $x = 3x - 4$, then to get $f(2)$, you have to replace $x$ with 2." She writes $f(x) = 3x - 4$ *so* $f(2) = 3(2) - 4$, *so* $f(2) = 6 - 4 = 2$.

Sarah and Andy go back to work and are able to correctly compute the other function values. Andy interjects, "I forgot that's how functions work. I was just thinking of it as multiplication!"

$$f(x) = 3x - 4$$
$$f(1) = 3(1) - 4 = -1$$
$$f(2) = 3(2) - 4 = 2$$
$$f(3) = 3(3) - 4 = 5$$
$$f(4) = 3(4) - 4 = 8$$
$$f(5) = 3(5) - 4 = 11$$

Next, the teacher asks the students to compare $f(2+3)$ to $f(2)+f(3)$. Andy and Sarah volunteer to go to the board and work on the problem under this teacher's guidance while the rest of the class works in groups. They write:

$$f(2+3) \overset{f(5)}{=} 3(5) - 4 = 11$$

$$f(2) + f(3) = 3(2) - 4 + 3(3) - 4$$

$$= 6 - 4 + 9 - 4$$

$$= 7$$

$$f(2+3) \neq f(2) + f(3)$$

Although many of the students nod in agreement (much to the satisfaction of Sarah and Andy), Mitchell asks, "How did you get that? I used the distributive property and got $f(2+3) = f(2) + f(3)$." The teacher uses this opportunity to reiterate the issue, knowing that other students in past classes have had similar misconceptions. "The distributive property applies to products of numbers. It's true that functions often involve numbers and the notation looks similar to multiplication notation." He writes f(2) on the board and continues, "This notation is read "eff of 2" and means the output of the function $f(x)$ when $x = 2$."

A number of other students are appreciative of Mitchell's question, because they have the same misconception. After a little more class discussion, the difficulties are cleared up (for now) and students continue with the class activity. The teacher then writes *Does* f(3 – 2) *equal* f(3) – f(2)? on the board and Mitchell asks to share his answer with the class. He writes:

DOES $f(3-2) = f(3) - f(2)$?

$$f(x) = 3x - 4$$

$$f(3-2) = f(1) = 3(1) - 4$$

$$= -1$$

$$f(3) - f(2) = 3(3) - 4 - (3(2) - 4)$$

$$= 9 - 4 - (6 - 4)$$

$$= 5 - 2 = 3$$

No!!!

His classmates and teacher have a good laugh; then they go on to compute and simplify $f(a + b)$. Compound inputs are often problematic, but the students make use of the earlier discussion and help each other determine that $f(a + b) = 3(a + b) - 4 = 3a + 3b - 4$. Next, the teacher asks, "Does $f(a+b)$ equal $f(a)+ f(b)$?" Abby quickly writes:

$$f(a)+f(b) = 3a-4+3b-4 = 3a+3b-8$$

and announces, "They don't look the same!"

Sarah says, "Wait a second! We already showed that $f(2 + 3)$ doesn't equal $f(2) + f(3)$, so they're not equal when $a = 2$ and $b = 3$." The teacher replies, "That shows us that $f(a + b)$ does not always equal $f(a) + f(b)$. How can we use Abby's work to help us decide if they're ever equal?"

Andy suggests, "I don't think it's *ever* possible." He shares his work.

If $f(a+b)=f(a)+f(b)$

then $3a+3b-4 = 3a-4+3b-4$

$3a+3b-4 = 3a+3b-8$

$4=8$ ???

He continues, "We can subtract $3a$ and $3b$ from both sides of the equation. That means that $4 = 8$, but that's impossible, so $f(a + b)$ can never equal $f(a) + f(b)$!"

Next, this teacher presents the question, "Does $f(-a)$ ever equal $-f(a)$?" As the students begin to work, some choose specific values of $a$. Sarah writes:

$f(-1) = 3(-1) - 4 = -3 - 4 = -7$

$f(1) = -1$

so $f(-1) \neq -f(1)$

Several other students try some other values but become frustrated. Andy throws up his hands and says, "It must never work. We can't find any value for $a$ that works!" Sarah replies, "But we can never try every possibility for $a$. Maybe $a$ is a fraction or an irrational number, not an integer. Let's do what Abby and Andy did. If they're equal, $-4 = 4$, so $f(-a)$ never equals $-f(a)$."

$f(-a) = -3a - 4$

$-f(a) = -(3a-4) = -3a+4$

never equal!

This teacher then poses, "Does $f(-a)$ ever equal $f(a)$?"

Although some suggest that it can never happen, just like the other problems, Mitchell insists on trying and writes: *if $f(-a) = f(a)$, then $-3a - 4 = 3a - 4$, so $-3a = 3a$, which means $-3 = 3$!*

Sarah says, "Wait a second. To get $-3 = 3$, you have to divide both sides by $a$, but you can't do that when $a = 0$. Because $-0$ equals 0, it must also be true that $f(-0)$ equals $f(0)$."

Abby adds, "That's right, because $-3a = 3a$, then we can add $3a$ to both sides to get $0 = 6a$, and therefore $a = 0$! That means $f(a)$ *can* equal $f(-a)$, but only if $a = 0$."

This teacher then assigns the *Working with Functions, Part 2* on page A3-3 in the appendix.

## Meeting Individual Needs

For students who need more help with understanding functions, you might assign the *Am I a Function?* reproducible on page A3-4 in the appendix. Students cut out cards with functions on them and organize the functions into columns labeled *Function* and *Not a Function*. Students explain why they sorted the examples the way they did.

For students who are still struggling with "plugging in" to functions, it might be useful to try a few more examples with numerical inputs. When evaluating an input of the function $f(x)$ given by an algebraic formula, students must replace $x$ with that input everywhere they see $x$ in the formula. For example, if $f(x) = 3x - 4$, which is three times $x$ minus four, then

$f(7)$ is $3(7) - 4$, or three times seven minus four.

$f(12)$ is $3(12) - 4$, or three times twelve minus four.

$f(\text{BLOB})$ is $3(\text{BLOB}) - 4$, or three times BLOB minus four.

Thinking of "BLOBS" can help when computing things like $f(a + b)$, because whatever is in the parentheses (the input) is the BLOB, so $f(a + b) = 3(a + b) - 4$

## Additional Reading/Resources


Edwards, Michael Todd, and Jennifer Nickell. 2014. "Teaching Students About Functions with Dynagraphs." *Mathematics Teaching in the Middle School*. August 16. http://www.nctm.org/publications/mathematics-teaching-in-middle-school/blog/teaching-students-about-functions-with-dynagraphs/.

Nabb, Keith. 2010. "The Back Page: My Favorite Lesson: Functions, Functions Everywhere." *The Mathematics Teacher* 104 (3): 240.

# 24. Number Patterns into Functions

**DOMAIN:** **Functions**

**STANDARDS:** **F-IF.1.** Understand that a function from one set (called the domain) to another set (called the range) assigns to each element of the domain exactly one element of the range. If $f$ is a function and $x$ is an element of its domain, then $f(x)$ denotes the output of $f$ corresponding to the input $x$. The graph of $f$ is the graph of the equation $y = f(x)$.

**F-IF.4.** For a function that models a relationship between two quantities, interpret key features of graphs and tables in terms of the quantities, and sketch graphs showing key features given a verbal description of the relationship.

## Potential Challenges and Misconceptions

Many students do not realize that square numbers, triangular numbers, rectangular numbers, and pentagonal numbers actually represent the number's shape when modeled with centimeter cubes, dots, or *X*s. In addition, when the total number of factors is represented on a line plot, students are able to see that the square numbers have an odd number of factors. The more time students spend playing with numbers, the more likely they will see the patterns that become visual, such as star numbers, happy numbers, and the Fibonacci sequence.

## In the Classroom

Functions often represent "real-world" contexts, but can sometimes be extended to other contexts. In this module, students work on the border of these two worlds by starting with a function that represents a visually defined number sequence and then expand the function beyond counting number inputs.

This teacher begins the class by projecting the following set of dots:

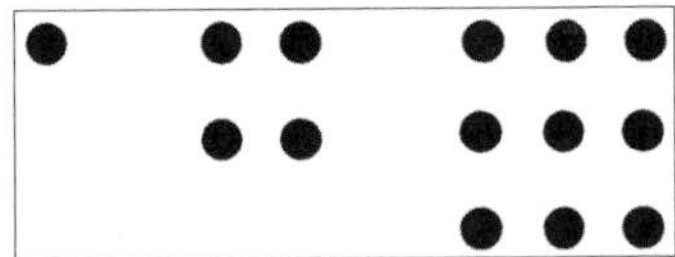

She asks her students how they might *see* the fourth and fifth square number. Gillian responds, "I see 4(four) rows of four dots for sixteen and the fifth square number, twenty-five, as five rows of five dots."

From this their teacher asks, "What are the tenth and fifteenth square numbers?" Many students start drawing rows of dots. Maria thinks for a moment and announces, "I imagined drawing five rows of five dots for the fifth square number and realized that the tenth square number would be 10 × 10 dots, so the tenth square number is one hundred. I did the same thing for the fifteenth square number and figured there would be 15 × 15, or 225 dots." Teneesha responds, "Yeah, that makes sense! That means the *n*th square number is $n \times n$, which equals $n^2$."

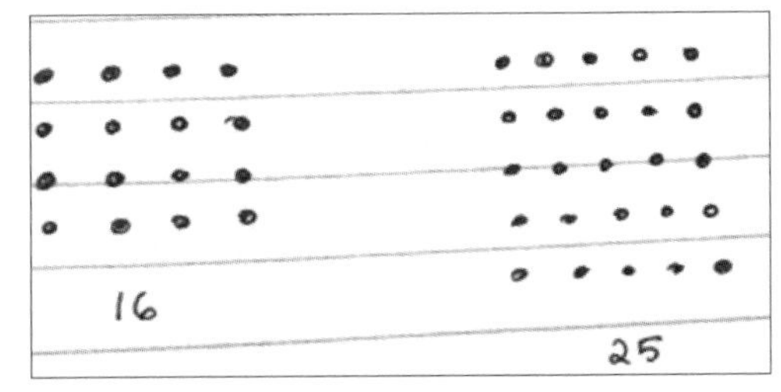




SECTION 3: FUNCTIONS

Next, this teacher asks how they might write this information as a function. Colin volunteers, "We can write it as eff of *x*." This teacher writes $S(n) = n^2$ and reminds her students that it is really helpful to use variables that relate to context [so in this case, $S(n)$ represents the *n*th square number].

This teacher then extends the situation and asks her students how they might evaluate the function $S(n) = n^2$ if the inputs are not counting numbers, for instance, $S(1.5)$ or $S(-3)$. Marcus asks, "Isn't negative three squared equal to negative nine?" Marcus raises a question that typically haunts students.

Teneesha goes to the document camera and writes and explains that $-3^2 = -9$ and $(-3)^2 = 9$. "But $S(-3)$ equals $-3 \times -3$, which is the square of $-3$; the $-3$ needs to be in parentheses." This teacher notes that Teneesha is demonstrating use of SMP 6: Attend to precision.

This teacher then asks, "Can anyone find a number *x* so that $S(x)$ equals 30.25? Can you find two numbers that work? Are there always two inputs for any output?" Teneesha starts writing, then grabs her calculator and announces, "It's 5.5 or −5.5!" as she finishes writing her solution.

$$x^2 = 30.25$$
$$x = \pm\sqrt{30.25}$$
$$x = 5.5 \text{ or } -5.5$$

Joey asks, "There will always be two answers for this function, right? You'll always have a plus or minus in the answer." Teneesha adds, "What about zero? Minus zero is still zero, so there's just one number whose square is zero. But all the others will have two answers."

This teacher asks the students to work in pairs on the reproducible *Double-Wides* on page A3-5 in the appendix.

## Meeting Individual Needs

Although some students are more than ready to work on the challenging double-wides problem, it is helpful for those students who are still processing this idea of function together with figurate numbers to work with students on the reproducible *Square Numbers* on page A3-6 in the appendix.

## Additional Reading/Resources

Wiesman, Jeff L. 2015. "Enhancing Students' Understanding of Square Roots." *Teaching Mathematics in the Middle School* 20 (9): 556–558.

# 25. Function Translations

**DOMAIN:** **Functions**

**STANDARDS:** **F-LE.5.** Interpret the parameters in a linear or exponential function in terms of a context.

**F-LE.1b.** Recognize situations in which one quantity changes at a constant rate per unit interval relative to another.

## Potential Challenges and Misconceptions

Early on in their dealings with functions, students often confuse input and output both symbolically and contextually. Although the symbolic misconceptions tend to work themselves out pretty quickly, facility with moving between symbolic and verbal or contextual representations comes gradually, in fits and starts. Another difficulty arises when students don't pay attention to the units of the inputs or outputs of the function. It is important to give students some time to think deeply about the connections between symbolic and verbal/contextual representations and to attend to units.

## In the Classroom

Before giving any instruction, one teacher begins his class by assigning the reproducible *Translate This* on page A3-7 in the appendix. Students work in triads. As the students are working, the teacher hears a discussion in the back of the room. The handout concerns some scientific instruments that have been dropped from a height of 90,000 feet (about 17 miles). The students are talking about problem 1, which asks them to translate symbolic statements into English and to evaluate their "truth." Ari claims, "The equation $f(0) = 90{,}000$ means that the instruments splash down after 90,000 seconds," but Zoe disagrees and says, "The input is the time since the instruments started falling, and that's 0 seconds. The output is the number of feet above the ocean the instruments are at that time. That means that the height of the instruments is 90,000 feet when the time is 0, but I'm not sure what a time of 0 means."

"That's no seconds!" Gloria adds, "I think 0 seconds just means the time the instruments were released from the balloon. And the information said that the instruments were released at a height of 90,000 feet, so that makes the statement true." The students discuss this a little more, and this teacher is happy that they are attending to units, because he knows units can be a trouble spot for a lot of students. This teacher is delighted as he listens to his students model SMP 3: Construct viable arguments and critique the reasoning of others.

This teacher overhears another group of students and is concerned that they are not paying attention to units, possibly due to their "need" to finish the activity first. Looking at part b of the question, Suzy asks, "How can we know if eff of seven is greater than eff of fifteen? We don't even know the formula!" Marcus responds, "The input is time and the output is height, so it means the height at time seven is greater than the height at time fifteen, but I don't know if it's true."

Mei has been thinking about this issue and thinks she has a solution. "The time is the number of seconds since the instruments were dropped from the balloon, and the height is the number of feet above the ocean. So it's saying that the height of the instruments 7 seconds after they're dropped is greater than the height of the instruments 15 seconds after they're dropped. And part c means the instruments are 95,000 feet above the ocean after 15 seconds."

Marcus is listening intently and suddenly responds, "The instruments are going to fall once they're dropped, so they'll fall farther in 15 seconds than they will in 7 seconds, so the height is greater after 7 seconds. And $f(15)$ can't possibly be 95,000 because that would mean the instruments were 95,000 feet above the ocean 15 seconds after being released from the balloon. That's 5,000 feet higher than when they were released. That's impossible!"

Mei adds, referring to part d, "I don't think $f(15)$ can equal 0, because that would mean the height of the instruments would be 0 feet above sea level after 15 seconds. Could they really fall 90,000 feet in 15 seconds? It seems like it would take a lot longer than that. I saw some skydivers on TV once, and they took about 10 seconds to open their parachute, and they jumped out of the plane less than a mile in the air."

Suzy answers, "I think you're right, but could $f(90{,}000)$ equal 0? Ninety thousand seconds seems like a long time!" Marcus grabs a calculator, punches a lot of buttons, and says, "Look, 90,000 seconds is 1,500 minutes, which is 25 hours! I don't think it would take a day to fall 90,000 feet, but I'm not sure! I *do* know that $f(14)$ can't equal $f(15)$, though. The height of the instruments won't be the same after 15 seconds as it was after 14 seconds, unless they have some magical way to hover!"

A third group is already discussing problem 2 on the handout, which asks them to translate English phrases into function notation. According to Joey, "The height of the instruments after 3 minutes is $f(3)$, right? The time is 3." At first, the teacher thinks the rest of Joey's group is going along with his statement, because they all nod. But just before the teacher intervenes, Carol suggests, "I don't think $f(3)$ is right. We want 3 minutes, but time is measured in seconds for this problem. Because 3 minutes is 180 seconds, shouldn't the answer be $f(180)$?" Joey quickly replies, "Oh, that's right. Thanks, Carol!"

The teacher is glad the students worked out the misconception on their own and makes sure to listen to other groups to make sure those students are correctly dealing with the units in problem 2. Just as he is moving to listen in on another group, he hears Carol, with renewed confidence, say, "Let's move on to the next one. Let's see, we need to determine the average velocity of the instruments from the time they were released until 5 seconds later."

As the teacher reaches the next group, Samara is showing her group her work on the average velocity translation. She is adding some labels to her work in response to questions from her table partners.

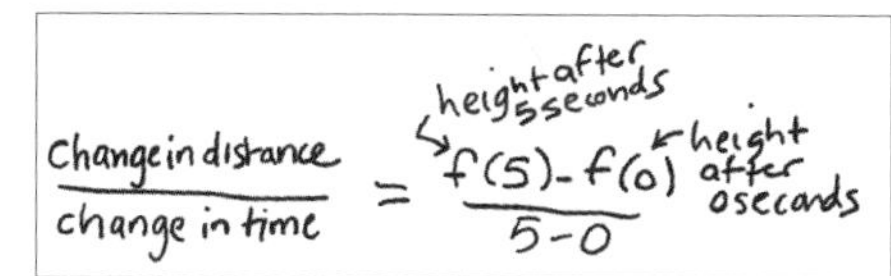


Samara is excited to see that problem 3 included a formula for the height of the instruments after $x$ seconds of falling: "The height $x$ seconds after being released from the balloon is $90{,}000 - 16x^2$ feet. For part a, the time is 15 seconds, so that means the instruments were 86,400 feet high after 15 seconds."

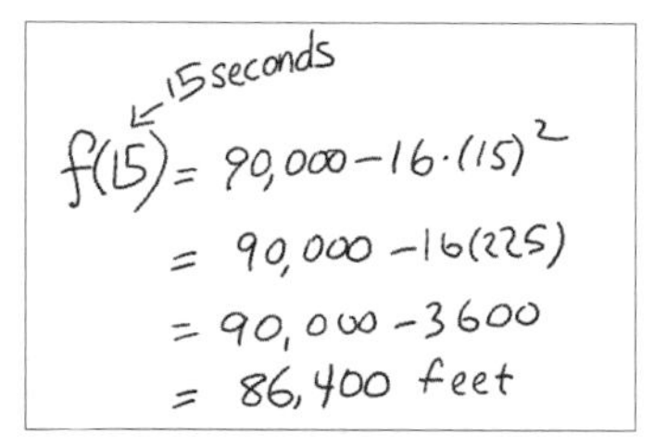


Toby adds, "Yeah, and after 3 minutes, which is 180 seconds, the height is 57,600 feet. That's almost 11 miles!"

$f(180) = 90{,}000 - 16 \cdot 180^2$

$180^2 = 32{,}400 \quad 90{,}000 - 32{,}400 = 57{,}600$

height after 3 minutes is 57,600 feet.

almost 11 miles

"I got the average velocity after 5 seconds to be 17,920 feet per second. That seems way too fast!" Samara can't figure out what she has done wrong.

Toby takes a look at Samara's work and says, "Oh, I see. You said the height after 0 seconds was 0 feet, but it's actually 90,000 feet. Here's what I got."

After asking his students to share their work, this teacher asks them if it makes sense to have a negative velocity. Some students looked puzzled until Johanna said, "Of course it makes sense because the height is decreasing as the time is increasing. I think the average velocity has to be negative as well." This teacher then assigns the handout *Translate That* on page A3-9 in the appendix.

$\frac{\text{change in distance}}{\text{change in time}} = \frac{f(5) - f(0)}{5 - 0}$ (height after 5 seconds; height after 0 seconds)

$\frac{\text{feet}}{\text{seconds}} \quad \frac{89{,}600 - 0}{5 - 0} = 17{,}920 \ \frac{\text{feet}}{\text{sec}}$

$f(5) = 90000 - 16(25) = 90000 - 400 = 89600$

$\frac{f(5) - f(0)}{5 - 0} = \frac{89{,}600 - 90{,}000}{5 - 0} = \frac{-400}{5} = -80 \ \frac{\text{feet}}{\text{sec}}$

## Meeting Individual Needs

It is helpful for students who struggle working with the *Translate This* questions to graph the ordered pairs that make up the function. Because of the dramatically different scales of domain and range in this problem, the two axes will need different scales, and students will likely need help in determining the scales. Graphs give a visual representation that often helps students understand the domain and range of the function, especially when the domain is restricted due to the context of the function. It is also extremely helpful for students to play around with various functions to experience how they act.

## Additional Reading/Resources


Erlina. 2012. "What Is an Inverse Function?" *Mathematics for Teaching*. February 26. http://math4teaching.com/2012/02/26/what-is-an-inverse-function/.

Hartter, Beverly. 2009. "A Function or Not a Function? That Is the Question." *The Mathematics Teacher* 103 (3): 200–205.

# 26. Celsius to Fahrenheit and Back Again

**DOMAIN: Functions**

**STANDARDS:** **8.F.1.** Understand that a function is a rule that assigns to each input exactly one output.

**8.F.4.** Construct a function to model relationships between quantities.

**F-BF** 1b. Combine standard function types using arithmetic operations.

## Potential Challenges and Misconceptions

Many travelers struggle to make the conversion of temperature from Celsius to Fahrenheit while traveling abroad because they haven't had much experience with the Celsius scale, and they don't remember the function that relates the two temperature scales. As when confronted with other unfamiliar metric measurements, many students struggle to make sense of having a body temperature of 37°C or hearing that the forecast high temperature will be 30°C on a summer day. Although the old adage says that familiarity breeds contempt, in this situation, *un*familiarity often breeds confusion.

## In the Classroom

One teacher begins by asking the students what they know about the way in which different countries record temperatures. He does this to validate the experiences of his students who have emigrated from European, Asian, and African countries. He is impressed that his students talk about Celsius, Fahrenheit, and Kelvin scales. On further investigation, he discovers that those students who mention Kelvin have relatives who are scientists. He surveys his class to determine how many students talk about temperature in Fahrenheit and how many use Celsius. He discovers that two-thirds of his students use Fahrenheit and one-third use Celsius. This teacher states that there are basically two temperature scales used in the world for everyday and many scientific purposes: Fahrenheit and Celsius (or Centigrade). In the United States, most people use Fahrenheit, but most other countries in the world use Celsius.

He begins his lesson by providing small groups with a thermometer on which Fahrenheit and Celsius scales are noted, a bowl, ice, a carafe of hot water, and a pitcher of cold water. He instructs his students to investigate the temperatures of the ice (which quickly melts), the hot water, and the cold water and to record their readings in a table. Some students decided to test other items, including their own body temperatures, water from the bubbler, water in their backpacks, and the temperature of the milk in their lunches. Each group makes a table and endeavors to find an explicit relationship between the Fahrenheit readings and the Celsius readings. This teacher overhears some students saying, "Wow, imagine if my temp was 37° and I didn't say it was Celsius. Folks would think I was dead."

After allowing his students to experiment with the thermometer, this teacher asks his students what they can surmise about the temperatures in Fahrenheit compared with those in Celsius.

Jennie volunteers, "The relationship has something to do with 32°." Marshall adds, "If we start with Fahrenheit then we can use a factor of two then adjust for the fact that that is just a little too much."

This teacher then asks his students questions about the Fahrenheit and Celsius scales: "At what temperature does water boil or freeze in Fahrenheit and in Celsius?" Most students familiar with the Fahrenheit scale record 212° and 32° and those experienced with Celsius correctly state that water boils at 100°C and freezes at 0°C. Some students remark that they have done this in their science classes. This teacher prods his students to express the relationships between Celsius and Fahrenheit using functional notation. Students work in pairs and most write $F(0) = 32$ and $F(100) = 212$. Some students add the values they had measured to their tables. As he walks around the class stopping at various groups, he challenges students to compute $F(-10)$ and $F(25)$, find $C$ so that $F(C) = 95$, and express in a sentence what $F(20) = 68$ means (a temperature of 20°C is equivalent to 68°F). As he walks around, he asks them if they can tell what type function relates Celsius to Fahrenheit. He is pleased when most students state it is linear. This activity engages students in SMP7: Look for and make use of structure, and helps students immensely. Most make a table like this

| C | F |
|---|---|
| –10 | |
| 0 | 32 |
| 25 | |
| | 95 |
| 100 | 212 |

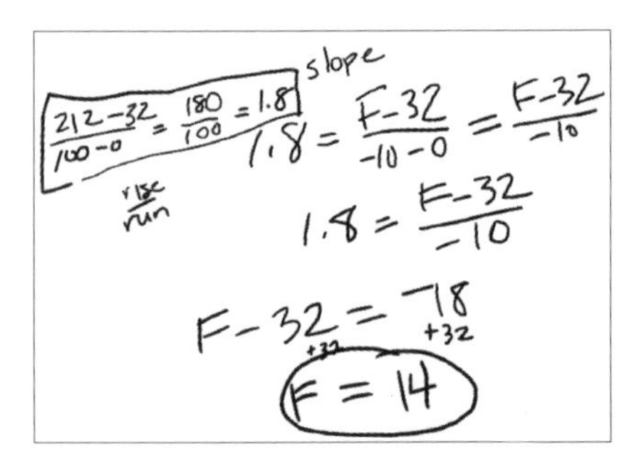


Kayla computes the slope between the points (0, 32) and (100, 212) and gets $\frac{180}{100} = 1.8$ or $\frac{9}{5}$. She then sets up an equation equating this value as the slope between (0, 32) and (–10, $F$) and gets $1.8 = \frac{F-32}{-10-0} = \frac{F-32}{-10}$. She concludes that $F - 32 = -18$, so $F = 14$.

Trevor says, "I thought of it like Kayla did, but I got a totally different answer. I got a slope of $\frac{5}{9}$, not $\frac{9}{5}$." Mickey replies, "I did the same thing, but then I realized that I computed the slope backward (or upside down). The slope is change in $y$ divided by change in $x$." Trevor replies, "Oh, yeah, that's right. I *always* make that mistake. Thanks, Mickey."

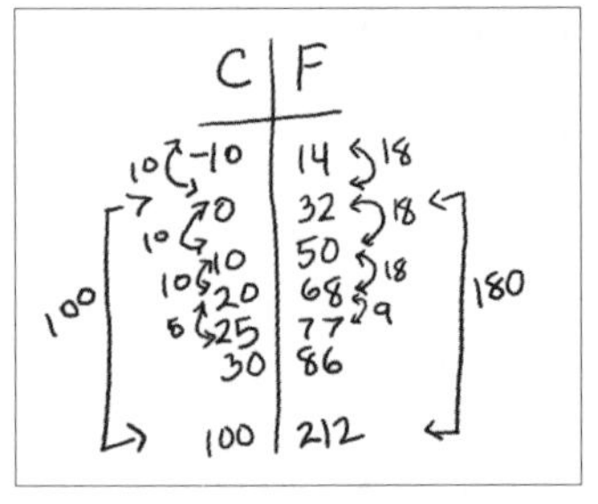


Jamal says, "I did something different from Kayla, but I got the same answer. When $C$ goes up by 100°, $F$ goes up by 180°, so if $C$ goes up (or down) by 10°, $F$ will go up (or down) by 18°."

"That means that when $C$ is –10°, $F$ will be 18° less than 32°, which is 14°. Thinking the same way, when $C$ goes up by 5, $F$ goes up by 9, so when $C$ goes up by 25 (or five 5s), $F$ must go up by five 9s, which is 45. So when the temperature is 25°C, the Fahrenheit temperature is 45 + 32, or 77°."

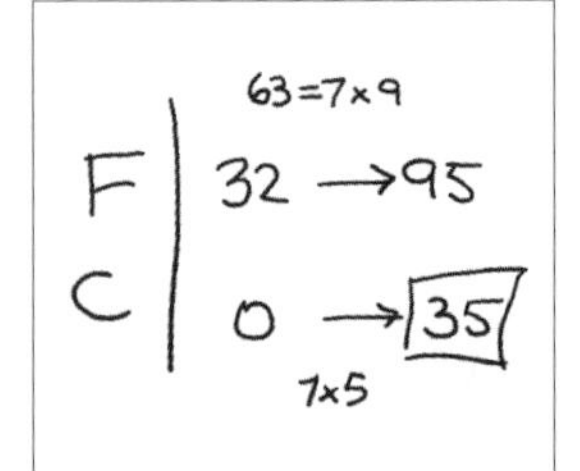


Michaela finds a way to use Jamal's reasoning to solve for the Celsius equivalent of 95°F. She says, "Using Jamal's method, when the Fahrenheit temperature goes up by 9°, the Celsius temperature goes up by 5°. To get from the freezing point, or 32°, to 95°, the Fahrenheit temperature goes up by 63°. Sixty-three is seven 9s, so the equivalent Celsius temperature must go up by seven 5s from the Celsius freezing point, 0. Seven 5s is 35, so 95°F is 35°C." Jamal looks over at her work and announces, "That's exactly what I did! Michaela is a mind reader!"

After agreeing that the completed table was correct as shown below, students determine the conversion function $F$ so that $F(C)$ is the Fahrenheit equivalent of $C°$ Celsius. Given the facts that the slope of this linear function is 1.8 and the "$F$-intercept" is 32, students recognize that $F(C) = 1.8C + 32$.

| C | F |
|---|---|
| -10 | **14** |
| 0 | 32 |
| 25 | **77** |
| **35** | 95 |
| 100 | 212 |

After completing this investigation, students are assigned the reproducible *How Cold Is It?* on page A3-10 in the appendix.

## Meeting Individual Needs

Because using a table to determine the slope of a linear function requires knowledge of how to find the slope given two points, some students may need to review that concept before working on these problems. It is also helpful for students to graph the two points they are given, (0, 32) and (100, 212), and use the line to find the missing values. For students who need a challenge, suggest they write a function to represent $C(F)$.

## Additional Reading/Resources

Hoyt, David, and Jeff Knurek. 2015. "Math Warm-Ups on a Cold Day." *Teaching Mathematics in the Middle School* 20 (6): 328–330.

# 27. Absolute Value Functions

**DOMAIN: Functions**

**STANDARD: F-1F.2.** Use function notation, evaluate functions for inputs in their domains, and interpret statements that use function notation in terms of a context.

## Potential Challenges and Misconceptions

Solving equations and inequalities involving absolute values can be difficult for many students, at least initially, but most will find ways to think about these problem types through experience. One common problem with compound expressions involving absolute values comes from the well-meaning, but often misleading, suggestion to translate every word individually into mathematical symbols. Younger children are often told, for example, that "*difference* means subtract." Ironically, when dealing with specific positive numbers, and not variables, subtracting lesser numbers from greater numbers cancels the first mistake out, because $a - b$ is the difference between $a$ and $b$ when $a > b$. This is not an issue with younger children, because they're typically dealing with positive numbers. However, an older student might be asked to determine the difference between –1 and –5. Because 5 is greater than 1, they might mistakenly compute –5 minus –1, and report that the difference between –1 and –5 is –4, rather than 4. When dealing with variables, we don't know which of the quantities is greater than the other, in general, so the "safe bet" is to use $|a - b|$ or $|b - a|$ for the difference of $a$ and $b$, because they're correct whether $a$ is larger than $b$ or not. That is, the difference of $a$ and $b$ might be $a - b$ or it might be $b - a$, depending on which of $a$ and $b$ is greater (i.e., the rightmost on the number line).

Working through the "story" slowly and deliberately, translating "chunks" instead of individual words, is often most productive. Getting better at solving absolute value equations and inequalities really does just take careful, concentrated practice.

## In the Classroom

One teacher, with these thoughts in mind, starts her lesson by writing English statements on the board and asking her students to solve and then rewrite the statements using mathematical symbols and variables. (These sentences are provided in the reproducible *Absolute Translations* on page A3-11 in the appendix.) Each student is able to rewrite *Five times a certain number is thirty-five* as $5(n) = 35$ or $5n = 35$. This teacher notices that some students use different letters than $n$ to represent the number. The most common letter used is $x$, which does not surprise this teacher, because $x$ is often used as a generic variable. The next three problems lead to some class disagreement and discussion.

For the phrase "The square of a number is sixteen," there is disagreement among the students. Toby says, "The second one is easy—the only number it can be is 4."

Tamara quickly responds, "Four works, but so does –4. A number can be negative, so that's a solution, too. And if I rewrite the sentence using mathematical symbols, I'd get $n^2 = 16$, and $n = 4$ and $n = -4$ are both solutions to that equation."

Toby thinks about it and tentatively adds, "I get that $n^2 = 16$ is the equation to solve, but when I solve it, I take the square root of both sides and get $n$ equals the square root of 16, which is 4. Why don't I get –4, too?"

Tamara replies, "Remember, the solutions to $n^2 = 16$ are plus or minus the square root of 16 because if you graph $y = x^2$, there are two places where the $y$-coordinate equals 16: $x = 4$ and $x = -4$."

Phung adds, "Oh yeah, remember in class we talked about this? The square of –3 is 9 and the square root of 9 is 3, so the square root of the square of –3 is 3." This teacher asks Phung to write that on the whiteboard. She records $\sqrt{(-3)^2} = \sqrt{9} = 3$.

Oscar raises his hand and adds, "It's like the absolute value function. Isn't that right? The square root of $x$ squared is the absolute value of $x$." He writes on the board:

$$\sqrt{x^2} = |x|$$

"That makes sense, because you get a nonnegative output whether $x$ is positive, negative, or zero. We saw that when we were working on piecewise functions! To solve $x^2 = 16$, we take the square root of both sides, which gives us the absolute value of $x$ equals 4, so $x$ is 4 or –4."

Brian adds, "I worked on the translation of that statement another way. I said the equation was $x^2 = 16$ but solved it by first saying that's the same as $x^2 - 16 = 0$. Then, because $x^2 - 16$ is the difference of two squares, it equals $x - 4$ times $x + 4$, so either $x - 4$ or $x + 4$ equals 0. That means, either $x$ equals –4 or $x$ equals 4." This teacher asks if anyone has anything different to share. When no one volunteers, she moves on to parts c and d, which state, "The difference between a number and five is three," and "The difference between five and a number is three."

As the class works on solving and translating these statements, this teacher hears interesting discussions. Sonya and Jane decide to solve all the problems, then go back to translate each of them. Jane says, "I got two for 'The difference between a number and five is three' and eight for 'The difference between five and a number is three,' but then I realized that both eight and two work on both of the problems, because the difference between eight and five is the same as the difference between five and eight. Difference is just how far apart the numbers are."

Oscar overhears them and leans over to say, "But I translated 'The difference between a number and five is three' as $n - 5 = 3$, and I translated 'The difference between five and a number is three' as $5 - n = 3$. The solution to the first equation is $n = 8$ and the solution to the second equation is $n = 2$."

Sonya replies, "But the difference can be three even if the number is less than five, so you really want to solve $n - 5 = \pm 3$. That's just like absolute value! I think the translations for both of them should be the absolute value of the quantity $n$ minus five is three" and wrote $|n - 5| = 3$ in her notebook.

This teacher asks Sonya, Jane, and Oscar to share their observations with the class, because several of their classmates have the same initial reaction as Oscar. This teacher then assigns *Mix and Match* on page A3-12 in the appendix.

SECTION 3: FUNCTIONS

## Meeting Individual Needs

Because it is often more difficult to write problems than it is to solve them, one way to engage students who have mastered translating from written expressions to mathematical expressions and vice versa is to challenge them to write the problems for their peers to solve. Ask them to think about situations in which it doesn't matter if something is positive or negative; what matters is how far from zero the number is or how far apart two quantities are, without caring which is greater than the other.

## Additional Reading/Resources

Beigie, Darin. 2014. "The Algebra Artist." *The Mathematics Teacher* 20 (5): 258–265.

Wei, Shiyuan (Steve). 2005. "Sharing Teaching Ideas: Solving Absolute Value Equations Algebraically and Geometrically." *The Mathematics Teacher* 99 (1): 72–74.

# 28. Modeling Exponentially

**DOMAIN:** **Functions**

**STANDARDS:** **F.8.5.** Describe qualitatively the functional relationship between two quantities by analyzing a graph.

**F-LE.1.** Distinguish between situations that can be modeled with linear functions and with exponential functions.

**F-LE.A.1a.** Prove that linear functions grow by equal differences over equal intervals and that exponential functions grow by equal factors over equal intervals.

## Potential Challenges and Misconceptions

Many students struggle to determine whether an input-output table represents an exponential function, because they cannot use finite differences but rather must compare ratios of the outputs. Most students have been told that to determine whether the data in a table are a linear or quadratic (or other polynomial) function, they must add or subtract the values in the output using the method of finite differences (if the input grows by consecutive integers). It is often difficult for students to change from thinking additively to thinking multiplicatively. Students need more practice with various situations that grow exponentially. Note that by convention, exponential functions are always of the form $f(x) = b^x$, but functions like $f(x) = Ab^x$ exhibit the same multiplicative growth, so we will include them under the umbrella of "exponential functions."

## In the Classroom

One teacher begins her class on exponential growth by reading the story *The King's Chessboard*. This story is about doubling the amount of rice placed on each of the sixty-four squares on a chessboard: one grain of rice on the first square, two grains on the second square, four grains on the third, and so on. Once students understand the situation, this teacher asks them to predict how many grains of rice will be on the sixty-fourth square. To engage students in SMP2: Reason abstractly and quantitatively, she asks them to think carefully about whether their predictions are reasonable, given the information so far, and she has them write their predictions on small sticky notes, which she then collects. This teacher quickly scans the predictions and tells the class the range of predictions. Later, she will ask the students to do a statistical analysis of their predictions.

After reading and discussing the story, they come to the consensus that there will be $2^{63}$ grains of rice on the sixty-fourth square (between nine billion and ten billion). This teacher displays the following table and poses the question, "Could the function that represents the data in this table be an exponential function? Convince your shoulder partner that your response is correct."

| $x$ | $f(x)$ |
|---|---|
| 1 | 12 |
| 2 | 72 |
| 3 | 432 |

After several minutes, this teacher asks volunteers to share their thinking. Kareem offers, "We don't really have a lot of points to work with, so we can't be sure." Nina agrees, adding, "Yeah, but remember, we never can have enough points to know for sure. We'll never know what the next input-output pair is unless we know how the table was constructed." This teacher acknowledges this statement but reminds students, "Remember, I asked you whether the table could possibly represent an exponential function."

The students don't respond at first, so the teacher asked whether they remember the patterns of exponential functions. Sally asks, "Are these the ones where you multiply instead of add?"

Geoff calls out, "Wait, what? I think you might be right, but what do you mean by multiplying instead of adding?"

Sally replies, "Well, remember with linear functions, the difference between one output and the next is always the same, always constant if the input increases by one each time. Well, with exponential functions, when you divide consecutive outputs, you always get the same thing. That makes sense because exponents are like repeated multiplication."

Kareem looks at the table, sits up quickly, and says, "OK then. I get it. For this function, if we compute $f(2)$ divided by $f(1)$, we get 72 divided by 12, which is 6. Also, $f(3)$ divided by $f(2)$ is 432 divided by 72, which is also 6, so I guess this might be an exponential function. If it is, then $f(4)$ should be 432 times 6, which is, um, 2,592." He takes a moment to check his calculation and says, more confidently, "Right, $f(4)$ will be 2,592. And $f(5)$ will be that times 6..."

This teacher interrupts, saying "That's right, Kareem, but today we're not going to find all of the outputs. We're going to find a formula for $f(x)$ that fits the data in the table, and it should also match the output you got for $x = 4$."

This teacher next introduces the fact that exponential functions have formulas that look like $f(x) = Ab^x$, where $A$ and $b$ are numbers that can be determined by examining input-output pairs. She also reminds students of the written expressions they used in the *Taxes, Interest, and More* activity. She asks Aiden how he might think of these exponential functions. Aiden responds, "I knew this looked familiar. We took the starting value times the growth rate raised to the time period. So now we need to figure out the growth rate, right?" This teacher then asked her students to gather into their working groups and work on problems 2 and 3 in the *Which Exponential Function Is It?* reproducible on page A3-14 in the appendix.

After students work on the problems for a while, she asks each group to record their work on easel-sized paper and to hang the papers around the room. She wants her students to engage in SMP3: Construct viable arguments and critique the reasoning of others, through a facilitated gallery walk. Once all the posters are hanging, this teacher provides sticky notes at each piece of work. She tells the students to choose a representative from each group to write two comments for each poster, one that asks a question and one that compliments some part of the work being displayed. When students have completed their gallery walk and posted questions and compliments, the teacher facilitates a mathematical discussion. Josh's group asks Kareem, "So what's the formula for $f(x)$?"

Kareem replies, "We just computed $f(1)$ and $f(2)$ two ways and set them equal. That gave us two equations involving $A$ and $b$, so I solved for $A$ and substituted into the second equation and got $b = 6$ and $A = 2$." Josh repeats his question, "So, what is the equation for $f(x)$?"

$12 = f(1) = Ab$

$72 = f(2) = Ab^2$

$Ab = 12 \rightarrow b = \frac{12}{A}$ $\left(A = \frac{12}{b}\right)$

$Ab^2 = 72$

$\frac{12}{b} \cdot b^2 = 72$

$12b = 72$

$b = 6$

$Ab = 12$

$A \cdot 6 = 12 \rightarrow A = 2$

Kareem replies, "My bad. I'll write it up there. Oh, we got the same thing as Sally got." This teacher tells Kareem that he didn't have to write it now and reminds him to answer the question that is asked.

Sally reports out next. "We got the same equations as Kareem's group, but we did it another way. Nina suggested we could divide instead of substitute. When we divided, we got $Ab^2$ over $Ab$ equals $b$ and 72 divided by 12 equals 6, so $b = 6$. Then, because $Ab = 12$, $A$ times 6 equals 12. The function is $f(x) = 2(6^x)$. We checked, and $f(3)$ and $f(4)$ were what we expected." This teacher hears a lot of *ahas* when Sally shares her group's strategy, and she compliments the group on the work.

$72 = A \cdot b^2$

$12 = A \cdot b$

$\frac{72}{12} = \frac{A b^2}{A b}$

$6 = b$

$12 = A \cdot b$

$12 = A(6)$

$A = 2$

$y = 2 \cdot 6^x$

$f(x) = 2 \cdot 6^x$

$f(3) = 432$ ?

$f(3) = 2 \cdot 6^3$

$= 2 \cdot 216 = 432$

This teacher decides to have the class solve problem 3 as a large group, with Nina being a scribe at the board. She hadn't expected that students would recognize that $b$ could be 2 or $-2$ (with the method they used), but it led to a good observation and reinforced the concept that $n^2 = (-n)^2$, so the equation $b^2 = 4$ actually has two solutions. See the discussion in the "In the Classroom" section of "Absolute Value Functions" for more discussion of this important concept.

This teacher next asks her students to work in pairs and write one exponential problem that she will assign to the class to solve.

$g(x) = A \cdot b^x$

$12 = A \cdot b^4$

$3 = A \cdot b^2$

$\frac{12}{3} = \frac{A b^4}{A b^2}$

$4 = b^2$

$b = \pm 2$

to find $A$:

$3 = A \cdot (\pm 2)^2$

$3 = 4A$

$\frac{3}{4} = A$

$g(x) = \frac{3}{4}(\pm 2)^x$

$g(7) = \frac{3}{4}(\pm 2)^7$

$= \frac{3}{4} \cdot \pm 128 = \pm 96$

but $g(7) = 96$ so $b = 2$

## Meeting Individual Needs

Some students may require more scaffolding for the earlier problems that ask them to determine whether a function is exponential. You might suggest that they add a column to record consecutive *ratios*, rather than consecutive differences, because exponential functions have the property that the ratios of consecutive outputs are constant. This contrasts with linear functions, where consecutive outputs share common differences, or quadratic functions, where consecutive *second differences* are constant.

Also, for students who need more practice with exponential functions, you might assign *Coupon Stacks* from page A3-15 in the appendix. This activity gives students an opportunity to engage in SMP4: Model with mathematics. For students who need more of a challenge, you might assign the reproducible *Tea with Mrs. Wiley* on page A3-16 in the appendix to provide them with a glimpse of a use of an exponential function. Note that Newton's Law of Cooling usually involves the exponential function $e^x$ rather than $2^x$, as in this activity.

## Additional Reading/Resources

Birch, David. 1988. *The King's Chessboard.* New York: Penguin.

Wanko, Jeffrey. 2005. "Giving Exponential Functions a Fair Shake." *Teaching Mathematics in the Middle School* 11 (3): 118–124.

# 29. Falling to Pieces

**DOMAIN:** **Functions**

**STANDARD:** **F-IF.7b.** Graph square root, cube root, and piecewise-defined functions, including step functions and absolute value functions.

## Potential Challenges and Misconceptions

Many students have limited understanding of piecewise functions, mainly because they are overlooked in many mathematics curricula in the middle school. Because students have been working with linear, absolute value, quadratic, and exponential functions, they should have no great difficulty working with piecewise functions. The greatest misconception is generated by the restrictions placed on the domain of the function—that is, some students struggle to determine which "piece" of the function applies to a given input. With practice, all students can be successful in understanding these functions.

## In the Classroom

As this teacher usually does, she posts a range question to determine how much her students know about piecewise functions. She writes the following on the board and asks students to compute a variety of function values:

$$f(-27),\ f(-15),\ f(-4),\ f(0),\ f(3),\ f(8),\ \textit{and}\ f(124) \text{ if } f(x) = \begin{cases} -x \text{ if } x < 0 \\ x \text{ if } x \geq 0 \end{cases}$$

Many of the students struggle with the restrictions put into this function and express their concerns. Marty says, "I can't tell what any of the outputs should be. It says that $f(3)$ is −3 and it says that $f(3)$ is 3, but that's impossible because there can be only one output."

Sara responds, "I think $f(3)$ is supposed to be 3 because 3 is greater than or equal to 0. The formula says that $f(x)$ equals $x$ if $x$ is greater than or equal to zero. Because −4 is less than 0, we use the other part of the formula, so $f(-4)$ is minus −4, which is 4. We can also think of the minus sign as meaning 'the opposite of' and the opposite of −4 is 4."

Xavier chimes in, "I think I get it. Because −15 is less than 0, we use the top part of the formula, so $f(-15)$ is the opposite of −15, which is 15. But because 8 is greater than or equal to 0, $f(8)$ equals 8.

Marty then adds, "Wait a second. That means $f(-27)$ equals 27, $f(-15)$ equals 15, and $f(-4)$ equals 4, $f(0)$ equals 0, $f(8)$ equals 8, and $f(124)$ equals 124. That's just like absolute value!" This teacher then explains that the function is called a *piecewise function* because it is defined in pieces. To find a particular output, you first need to determine which piece applies to the given input.

This teacher then asks students to graph the function $y = f(x)$ using Geometer's Sketchpad. (This teacher uses Geometer's Sketchpad, but alternatives are available.) The graphs all look like one of the two shown here.

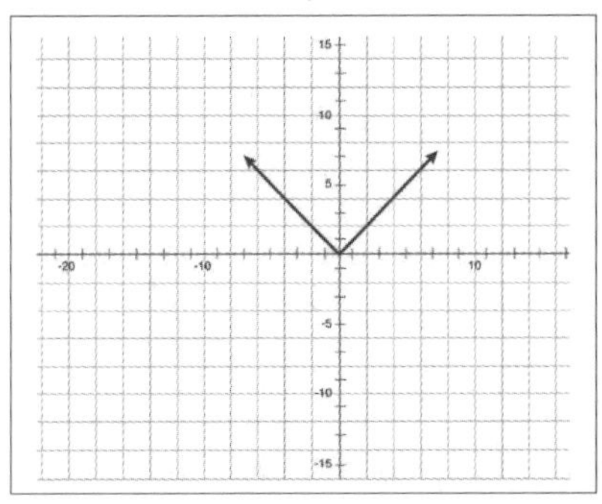

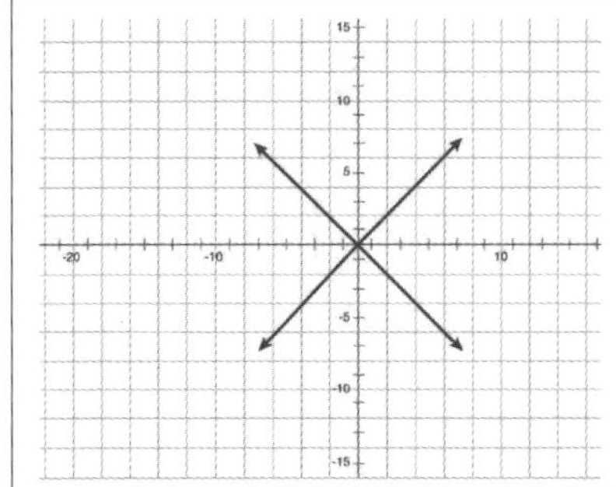

The class has a long discussion about which graph is correct. Everyone agrees that the graph of $y = f(x)$ should include $y = x$ and $y = -x$, but there is disagreement on whether the graph should include both lines. Marta points out, "The graph with two full lines isn't a function, so I think the correct graph is the one on the left. After all, we should only graph $y = x$ when $x$ is greater than or equal to 0 and graph $y = -x$ when $x$ is less than 0." Students come to consensus that the graph on the left is correct.

This teacher then asks the students to look at the graph of $y = g(x)$ and use the graph to compute the following values: g(−3), g(−1), g(0), g(1), g(1.5), g(3), g(4.7), g(10).

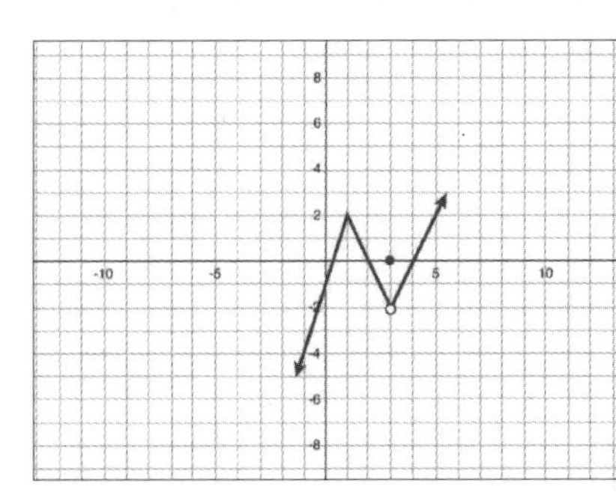

Determining g(−3) and g(10) leads to a lengthy discussion because the students cannot see the points corresponding to $x = -3$ or $x = 10$ on the graph. However, Marta points out, "Because the right and left parts of the graph have arrows, we should think of those rays continuing on forever to the left and right. The left part of the graph is part of the line $y = 3x - 1$, so $g(-3) = -10$ and the right part of the graph is part of the line $y = 2x - 8$, so $g(10) = 12$." This teacher finishes this part of the lesson by asking students to write a formula for g(x), which they agree is the one given here.

$$g(x) = \begin{cases} 3x - 1 & \text{if } x < 1 \\ 4 - 2x & \text{if } 1 \leq x \leq 3 \\ 1 & \text{if } x = 3 \\ 2x - 8 & \text{if } x > 3 \end{cases}$$

This teacher assigns the reproducible *Piecing Together the Cost* on page A3-18 in the appendix.

## Meeting Individual Needs

Some students will benefit from a review of how to graph various functions, in particular continuous and discrete functions, followed by graphing two functions on one set of axes. From there, they should have domain restrictions put on each function to allow them to see how piecewise functions operate. Students who struggle may benefit from working in a small group with the teacher as a facilitator or guide.

## Additional Reading/Resources

"Piecewise Functions Review." *CoolMath.com.* http://www.coolmath.com/precalculus-review-calculus-intro/precalculus-algebra/07-graphing-piecewise-defined-functions-01.

# 30. Why the Finite Differences Method Works

**DOMAIN:** **Functions**

**STANDARD:** **F-LE.1a.** Prove that linear functions grow by equal differences over equal intervals, and that exponential functions grow by equal factors over equal intervals.

## Potential Challenges and Misconceptions

So often students are provided with a set of procedures or rules to follow with little to no understanding of why and how they work and, more important, without understanding the mathematics behind those procedures and rules. In algebra, as students are challenged to determine the type of function a table represents, they are told to find the finite differences but are rarely involved in determining why this strategy works. More experiences, investigations, and applications for the use of finite differences are crucial if students are to connect this strategy with other mathematical connections to the importance of finite differences.

## In the Classroom

One teacher believes it is important that students understand the *limitation* of finite difference methods (for example, the methods only work when we assume that any perceived patterns in the differences continue to repeat or if we know that the underlying functions are polynomials). This teacher emphasizes the fact that any formula his students come up with is one among possibly many others. To help his students understand this concept, this teacher distributes the *Finite Differences Revisited 1* reproducible from page A3-20 in the appendix. This teacher poses the question, "If $f(x)$ is the function being modeled by the following table, what is $f(6)$?"

| $x$ | $f(x)$ |
|---|---|
| 2 | 7 |
| 3 | 11 |
| 4 | 15 |
| 5 | 19 |
| 6 | ? |

The students have a lot of experience with modeling linear functions, so they quickly determine that $f(6) = 23$, but this teacher is interested in digging a little deeper today. Toby says, "The outputs go up by four, so the next output after nineteen should be twenty-three."

This teacher probes further. "What kind of function does $f(x)$ seem to be? Can you find a formula for $f(x)$?" Zoe answers, "If the outputs keep going up by four when the input goes up by one, the function will be linear and the slope will be four." Joey responds, "I thought the $y$-intercept was four. Because $y$ is going up by four, the equation will be $y = mx + 4$." Zoe replies, "No, it's the slope—I'm positive!"

The class erupts into a loud but productive discussion, so this teacher calmly regains the attention of the class. "Let's see if we can work it out for ourselves. Let's suppose we have a linear function, $f(x) = Mx + B$." He writes the following table on the whiteboard and instructs the class to make a table of values for $x = 0, 1, 2, 3, 4$, and 5, asking each peer group to work on one of the $x$-values. In a few minutes, representatives of each group fill in the table:

| $x$ | $f(x)$ |
|---|---|
| 1 | $M(1) + B = M + B$ |
| 2 | $M(1) + B = M + B$ |
| 3 | $M(3) + B = 3M + B$ |
| 4 | $M(4) + B = 4M + B$ |
| 5 | $M(5) + B = 5M + B$ |

Next, this teacher adds a difference column to the table:

| $x$ | $f(x)$ | Difference: $f(x + 1) - f(x)$ |
|---|---|---|
| 1 | $M + B$ | $2M + B - (M + B) = M$ |
| 2 | $2M + B$ | $3M + B - (2M + B) = M$ |
| 3 | $3M + B$ | $4M + B - (3M + B) = M$ |
| 4 | $4M + B$ | $5M + B - (4M + B) = M$ |
| 5 | $5M + B$ | |

Zoe excitedly explains, "See? The difference is the slope of the line, like I said!"

Joey adds, "Cool, I see that now. But I see something else. Once you know what $M$ is, you can figure out $B$. Because we know $f(1) = M + B$, we get $B = f(1) - M$!"

Toby adds, "I have another way to figure out $B$. The $y$-intercept is the function's output when $x = 0$. Because the differences are always $M$, we can work up the table, instead of down. Because $f(0)$ will be $M$ less than $f(1)$, we get $f(0) = B$. Oh, that makes sense because using the formula, $f(0) = M \times 0 + B$, which equals $B$!"

This teacher replies, "So, if I understand you all correctly, the table shows that the difference column for a linear function is always the slope of the line and the $y$-intercept of the line is $f(0)$, or $f(1)$ minus the slope. Let's see if we can use that information to find a formula for the function represented by the next table."

Toby is quickly scribbling in her notebook. "I'm going to add a column for the differences."

| $x$ | $f(x)$ | $f(x + 1) - f(x)$ |
|---|---|---|
| 0 | -3 | |
| 1 | 3 | $9 - 3 = 6$ |
| 3 | 9 | $15 - 9 = 6$ |
| 4 | 15 | $21 - 9 = 6$ |
| 5 | 21 | $27 - 21 = 6$ |
| 6 | 27 | |

SECTION 3: FUNCTIONS

Toby continues, "It looks like it could be linear with the slope being 6, so $f(0)$ is 6 less than $f(1)$, which is −3. Let's see, $M = 6$ and $B = -3$, so the function's formula could be $f(x) = 6x - 3$."

Zoe replies, "I found the slope the same way, but I remembered that $f(1) = M + B$, so $3 = 6 + B$, which means that $B = -3$, so I got the same formula Toby got."

After discussing Toby and Zoe's work, the class solves the last problem in *Finite Differences Revisited 1*. The example in number 6 illustrates that the methods developed to predict function formulas are only predictions and that if you're given a table of values, you can never know how to fill in the next row unless you know something more about the function. For example, in the table above, we have no way of guaranteeing that $f(7)$ will equal twenty-eight even though it follows the pattern of differences that we noticed.

After discussing *Finite Differences Revisited 1*, this teacher assigns *Finite Differences Revisited 2 on* page A3-21 in the appendix.

## Meeting Individual Needs

For students who struggle, you might want to skip this activity, or skip number 5. For students who need a challenge, you might use all or part of the activity as an extension. However, the point of number 5 is to explain that there are many possible polynomials that fit a few input-output pairs. More importantly, there is no single answer to the question, "What is the next number in the sequence?"

## Additional Reading/Resources

Danielson, Christopher. 2014. "They'll Need It for Calculus." *Teaching Mathematics in the Middle School* 20 (5): 260–265.

# APPENDIX 1:
# Expressions

# Classroom Expressions

Name: ______________________________ Date: ____________________

Below are sample expressions you might use with the floor-sized model described in the module "Order of Operations."

1. $3 - 2(4 + 7)$
2. $7 + 5 \cdot 3 - 4$
3. $8 \div 4 \cdot 4 - 3 + 2$
4. $16 \div 8 \cdot 2 + 4 \cdot 3$
5. $3 + (6 - 2 \times 5)$
6. $6 - 3(1 - 3 - 4)$
7. $-2^3 - 4(-3 - 7)$
8. $6 + 3\ (8 - 3) \div 5 - 2^3$
9. $(-4)^3 - 7 \cdot \frac{1}{7} + 3$
10. $1 - 10(-10 + 9)^3$
11. $-7 - (-9) + 1(-1) \div 1$
12. $9.5 + (-6.1) - 2.8$

# In What Order?

Name: ______________________________ Date: ______________________

Write a series of equivalent expressions until you are left with one numerical expression. Record the operation you used next to each equivalent expression.

1. $6 - 8 + 3 \times 4$

2. $7 + 6(8 - 4 + 3) + 6 \cdot 3$

3. $3^2 + 4(2 - 5 \cdot 3) \div 14 - 12$

4. Gabriel says that the expression $-4 - 3(12 - 5) \div 3 \cdot -2^2$ is equivalent to the expression -32.

   Madison says that the expression $-4 - 3(12 - 5) \div 3 \cdot -2^2$ is equivalent to the expression 24.

   Who is correct? Justify your decision.

APPENDIX 1: EXPRESSIONS

# In What Order? Graphic Organizer

Name: ______________________________ Date: ______________________

| Operation | Expression |
|---|---|
| Parentheses | |
| Exponents | |
| Divide **OR** multiply from left to right | |
| Divide **OR** multiply from left to right | |
| Add **OR** subtract from left to right | |
| Add **OR** subtract from left to right | |

| Operation | Expression |
| --- | --- |
| Parentheses | |
| Exponents | |
| Divide **OR** multiply from left to right | |
| Divide **OR** multiply from left to right | |
| Add **OR** subtract from left to right | |
| Add **OR** subtract from left to right | |

# Equivalent Expressions

**Teacher:** Copy (on card stock if possible) and cut out each expression. Group cards by the letter at the top right corner: A cards, B cards, and C cards. Students place the equivalent expressions in each group to illustrate how they might use order of operations to simplify the original, shaded expression. The cards are arranged in the correct order from left to right here, so be sure to mix them up before giving them to students. Notice that there are incorrect as well as correct choices. Instruct students to record each step in a notebook or journal.

| | | |
|---|---|---|
| A. $-6 - 3(4 + 2 \times 7) \div 9 + 2^3$ (shaded) | A. $-6 - 3(4 + 14) \div 9 + 2^3$ | A. $-6 - 3(18) \div 9 + 2^3$ |
| A. $-6 - 3(18) \div 9 + 8$ | A. $-6 - 54 \div 9 + 8$ | A. $-6 - 6 + 8$ |
| A. $-12 + 8$ | A. $-4$ | A. $-9(18) \div 9 + 8$ |
| A. $0 + 8$ | A. $8$ | A. $-9(4 + 14) \div 9 + 2^3$ |
| B. $5 - 3(2 - 3 \times 4) \div 6 - 3^2$ (shaded) | B. $5 - 3(2 - 12) \div 6 - 3^2$ | B. $5 - 3(-10) \div 6 - 3^2$ |

| | | |
|---|---|---|
| B. $5 - 3(-10) \div 6 - 9$ | B. $5 + 30 \div 6 - 9$ | B. $5 + 5 - 9$ |
| B. $10 - 9$ | B. $1$ | B. $-2(2 - 3 \times 4) \div 6 - 3^2$ |
| B. $-2(-2 \times 4) \div 6 - 3^2$ | B. $5 - 3(2 - 3 \times 4) \div 6 + 8$ | B. $5 - 3(6 \times 4) \div 6 + 8$ |
| B. $2(6 \times 4) \div 6 + 8$ | B. $2(24) \div 6 + 8$ | B. $48 \div 6 + 8$ |
| B. $8 + 8$ | B. $16$ | B. $19$ |
| C. $3x - 2(x + 4) + 2x^2$ | C. $3x - 2x - 8 + 2x^2$ | C. $1x - 8 + 2x^2$ |
| C. $3x^2 + {}^-4$ | C. $1x - 4 + 2x^2$ | C. $3x^2 - 8$ |
| C. $11x^2$ | C. $3x - 2x - 4 + 2x^2$ | C. $5x - 8 + 2x^2$ |

# Greatest Prime Factor Graph

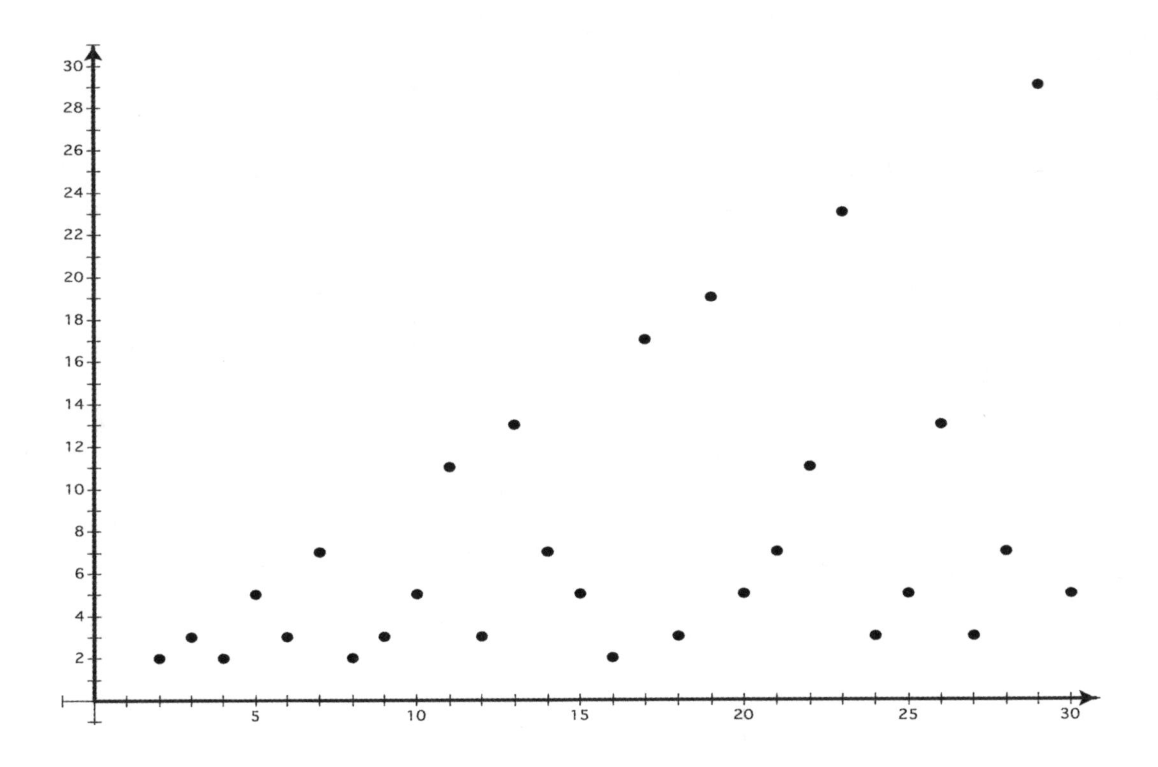

# What's My Expression?

Name: ______________________________ Date: ____________________

Write a verbal expression and a mathematical expression that represent the relationship shown in each table.

**1.**

| Input | 4 | 6 | 7 | 10 | $n$ |
|---|---|---|---|---|---|
| Output | 14 | 20 | 23 | 32 | |

**2.**

| Input | -5 | -2 | 3 | 5 | $n$ |
|---|---|---|---|---|---|
| Output | 32 | 14 | -16 | -28 | |

**3.**

| Input | -3 | -1 | 1 | 3 | $n$ |
|---|---|---|---|---|---|
| Output | -36 | -4 | -4 | -36 | |

**4.**

| Input | 0 | 1 | 2 | 3 | $n$ |
|---|---|---|---|---|---|
| Output | 1 | 3 | 8 | 11 | |

# Match the Equivalents

Name: ______________________________ Date: ______________________

Cut out each card. Match the verbal expression with its mathematical equivalent. Tape or glue the matching cards in your notebook

| | |
|---|---|
| **W**<br>**Add six to $n$, then multiply by two.** | **W**<br>**Divide the quantity $n$ plus six by two.** |
| **W**<br>**Square the quantity of $n$ plus four.** | **W**<br>**Add six to $n$, then square the sum.** |
| **W**<br>**Multiply $n$ by two, then add six.** | **W**<br>**Divide $n$ by two, then add six.** |
| **W**<br>**Square $n$, then add six.** | **W**<br>**Raise negative four to the second power.** |
| **W**<br>**Write the factors of $4n - 6$.** | **W**<br>**Multiply the product of four times $n$ by negative one.** |

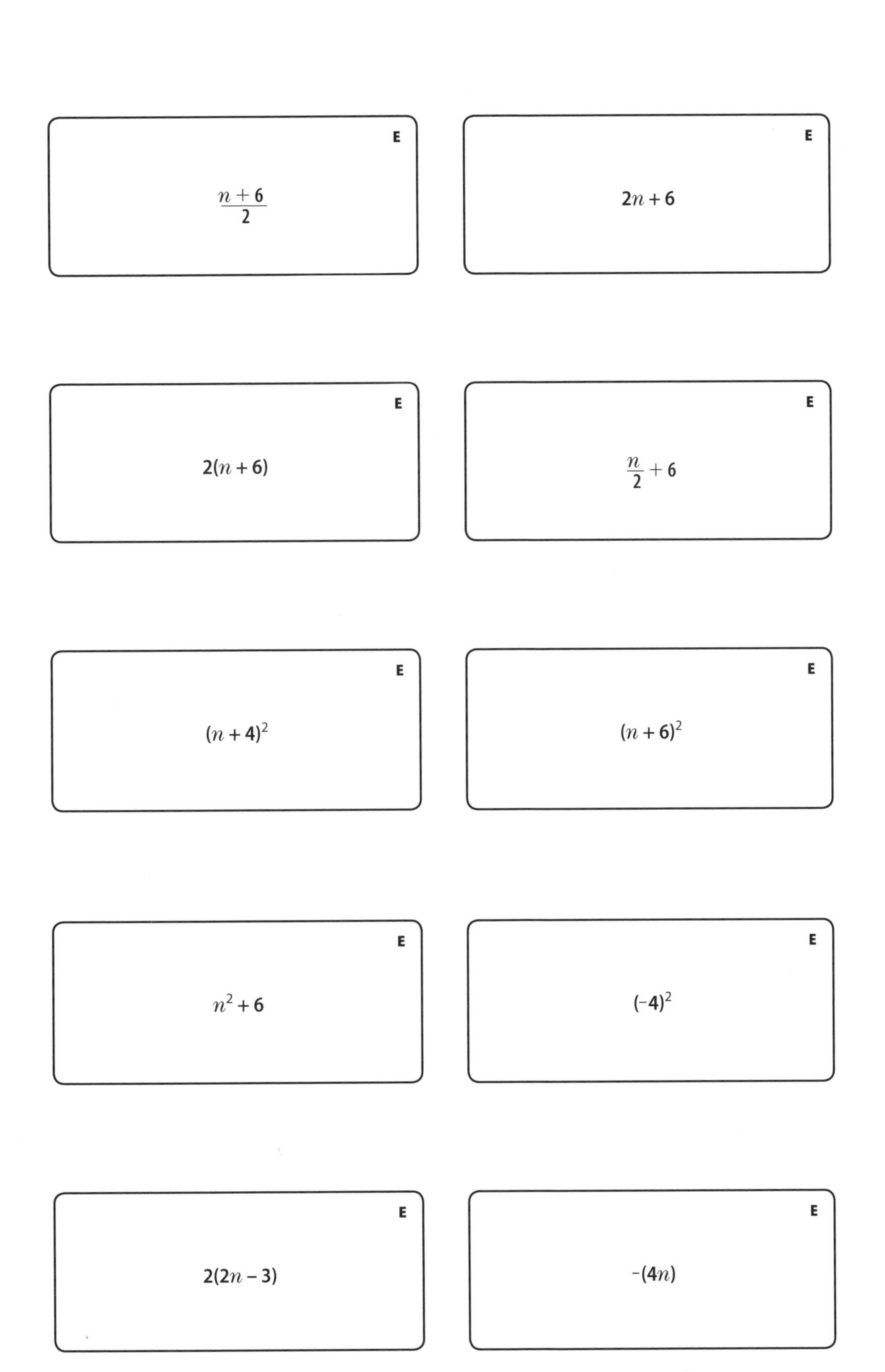
E
$\frac{n+6}{2}$
E
$2n+6$
E
$2(n+6)$
E
$\frac{n}{2}+6$
E
$(n+4)^2$
E
$(n+6)^2$
E
$n^2+6$
E
$(-4)^2$
E
$2(2n-3)$
E
$-(4n)$

# Growing Figures 1

Name: ______________________________ Date: ____________________

**Figure 1** **Figure 2** **Figure 3**

# What's My Rule?

Name: ______________________________ Date: ____________________

Examine the following figures.

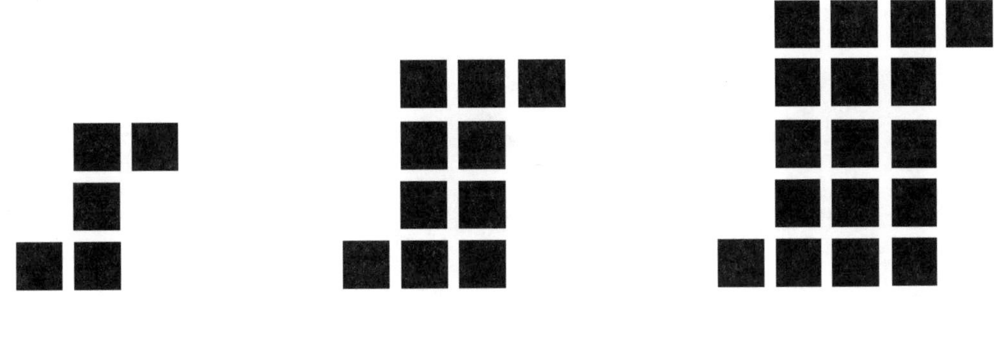


1. Describe what you see as the pattern grows.

2. Draw Figures 5 and 7.

3. Write an expression to find how many squares will be in the $n$th figure.

# Growing Figures 2

Name: ______________________________ Date: ____________________

For items 1–3, draw the fifth and seventh figure. Then write an expression to predict the total number of rhombi, stars, or squares in the $n$th figure.

1.

2.

3.

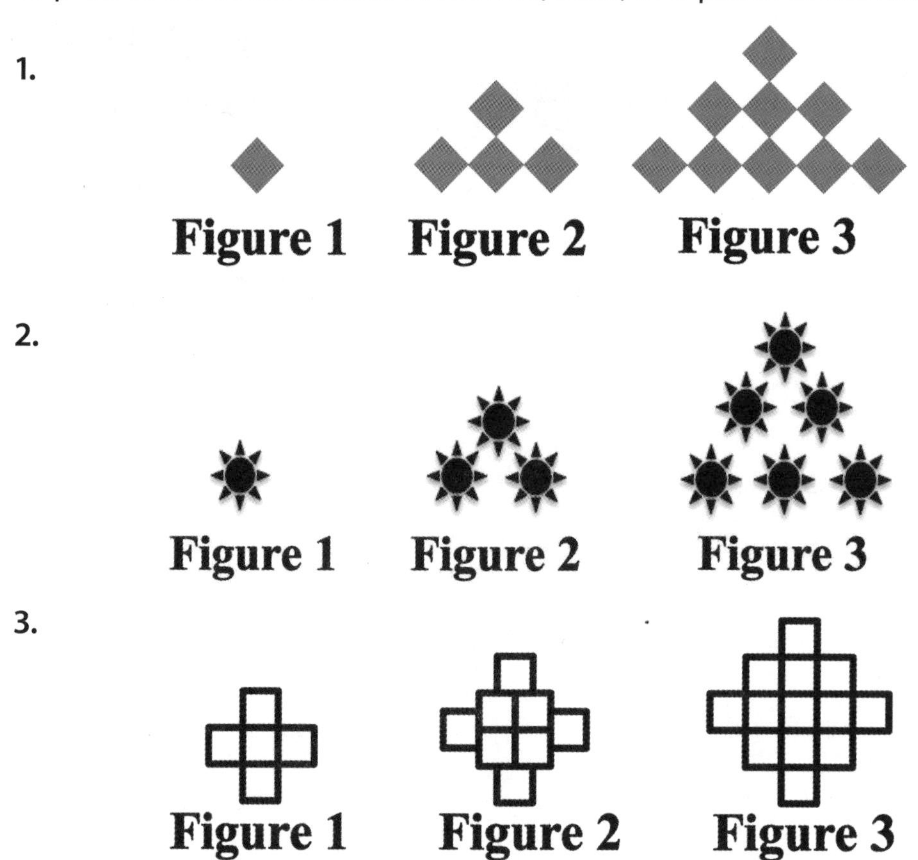


4. Draw the fifth and seventh figures. Assume that each edge of each triangle is 1. Find the perimeter of the figures. Write an expression to predict the perimeter of $n$th figure.

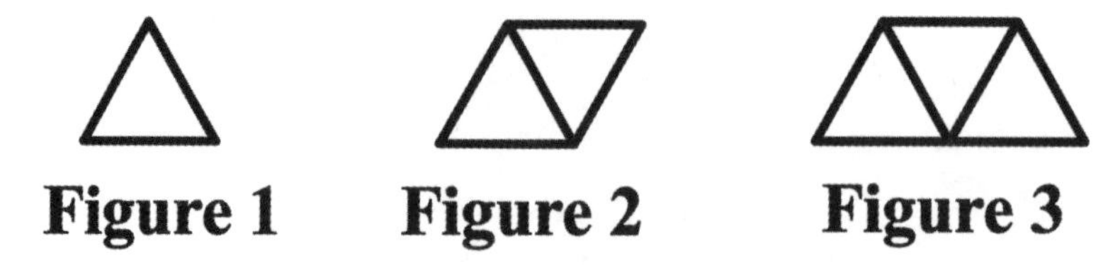

# My Counterpart





**Teacher:** Copy the following cards on card stock and laminate if possible. Students play *Concentration* as they try to make as many matches as possible. To do this, students work in pairs and deal the cards facedown. One student begins by turning over two cards. If the cards match, the player keeps both cards and takes another turn. If there is no match, the next player tries to get a match. The correct matches are side by side here, so be sure to mix the cards up before giving them to student pairs.

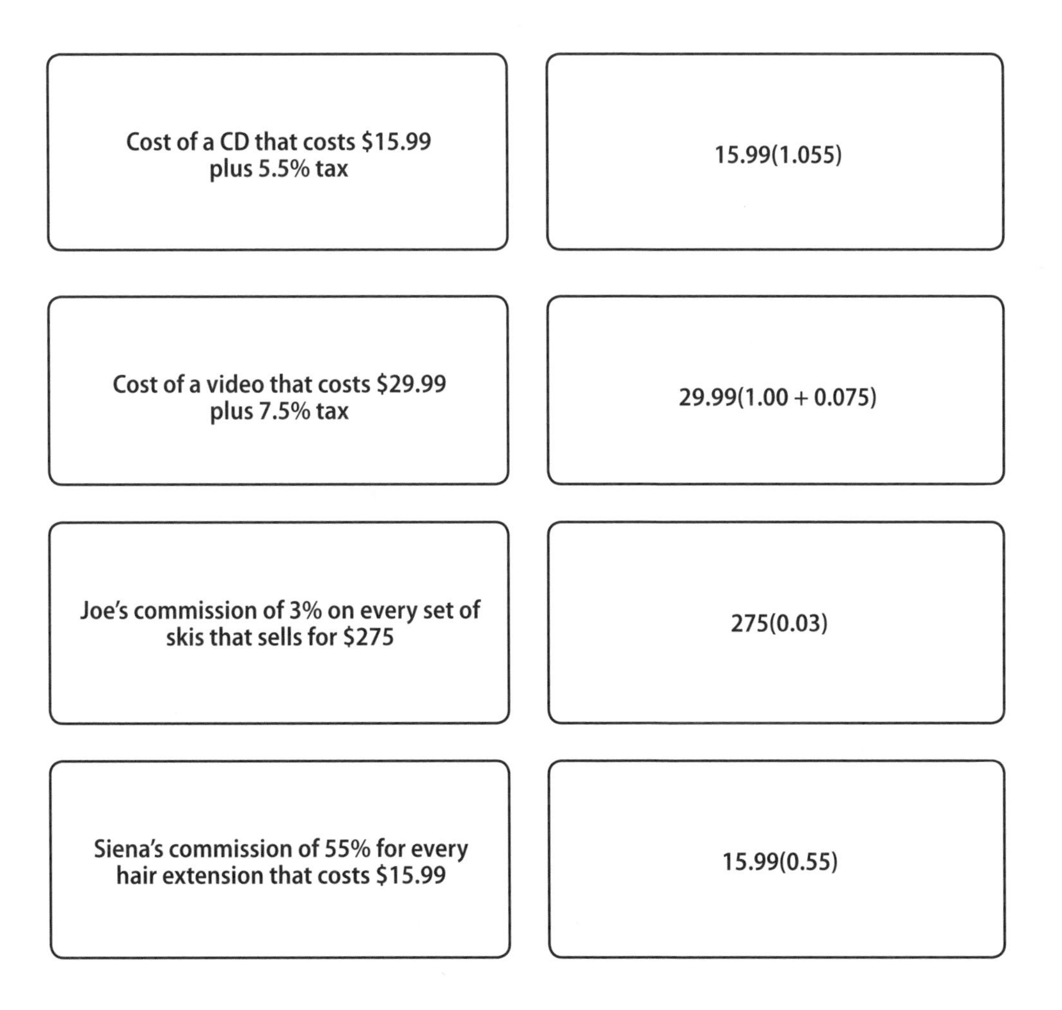

| | |
|---|---|
| Cost of a CD that costs $15.99 plus 5.5% tax | 15.99(1.055) |
| Cost of a video that costs $29.99 plus 7.5% tax | 29.99(1.00 + 0.075) |
| Joe's commission of 3% on every set of skis that sells for $275 | 275(0.03) |
| Siena's commission of 55% for every hair extension that costs $15.99 | 15.99(0.55) |

| | |
|---|---|
| **The cost of a $29.99 bag of beads that is on sale at 75% off** | **29.99(1.00 – 0.75)** |
| **Total repayment of a 5-year car loan of $21,000 with yearly interest of 3.5%** | **$21{,}000(1.035)^5$** |
| **The total cost of a dirt bike repair with 30% service fee on parts that cost $275** | **275(1.3)** |
| **Total cost of a bicycle that costs $725 plus 6.5% tax** | **725(1.00 + 0.065)** |
| **Total cost of a game that costs $7.25 plus 6.5% tax** | **7.25(1.065)** |

# How Much Will I Pay?

Name: ______________________________ Date: ____________________

1. Sam is planning to buy a racing bicycle that costs $900 for his son. He plans to pay for it over 2 years at a 6% interest rate compounded annually.

   a. Write an equation that will help Sam calculate how much he will pay for the bicycle at the end of the 2 years.

   b. Use your equation to find out how much Sam will pay for the bicycle after the 2 years.

2. Alex bought a television set for $850 and paid a sales tax of 8%. How much did he pay for the television set? Show your work on a tape diagram.

3. Because of budget cuts, the teachers in the Euclid School District were asked to take a 10% pay cut. The following year, the same district was able to give them a 10% raise. If the average pay per teacher is $55,000, how much money were they making after the pay cut followed by the pay raise?

4. A bookstore is selling books normally priced at $7 each, but today they're 25% off. If the store charges 6% tax, what is the final cost of one book?

# Taxes, Interest, and More

Name: ______________________________ Date: ____________________

Model each situation with a double number line or a table. Write a mathematical expression and solve the problem. Round your answers to the nearest cent.

1. Taylor's grandmother agrees to buy him a laptop if he agrees to pay the 7% sales tax. If the laptop costs $499, how much money will Taylor need?

2. Colton is saving up for a new skateboard that costs $119 plus tax. If the tax in his state is 5.5%, how much will the new skateboard cost?

3. Manny's dad works on commission at a car dealership. He typically receives 3% commission on each car he sells. He recently sold a luxury car for $88,400. How much of a commission did he earn?

4. Myrna deposited $6,500 in her bank. The bank pays 1.5% interest per year. How much money will Myrna have in the bank after three years?

# Venn Diagrams

Name: ______________________________ Date: ______________________

# Venn Diagram Representations

Name: ______________________________ Date: ______________________

Model and find the greatest common factor and least common multiple for the following.

**1.** $x^2y^3; x^3y^2$

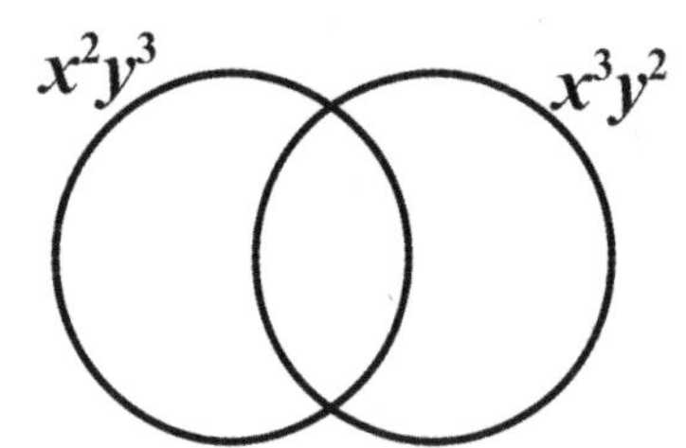


$a^4b^2; a^2b$

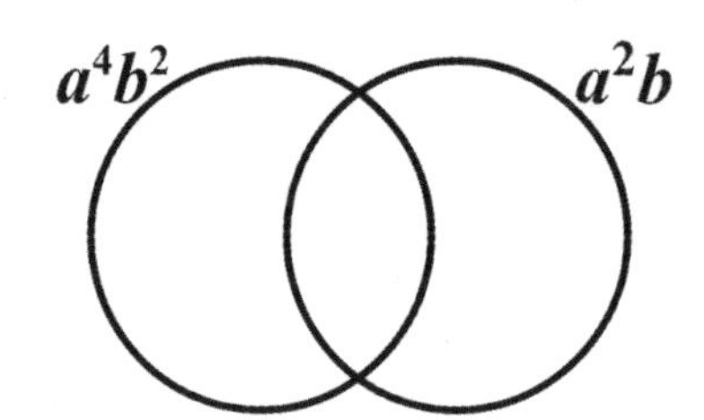


**2.** $c^5d^3; c^2d^2$

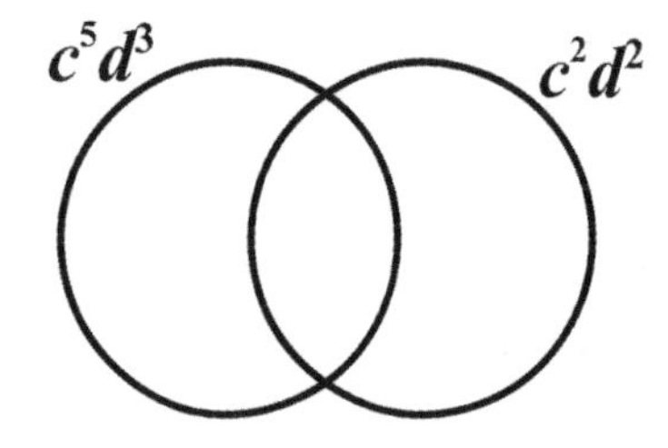


$x^4y^8; x^6y^2$

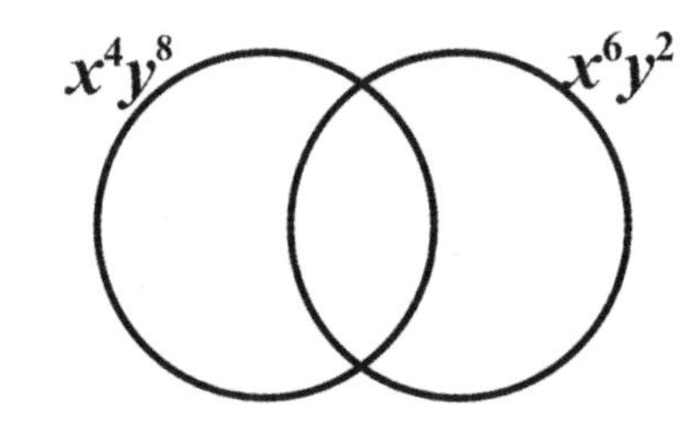


**3.** $a^3b^4; a^4b^5$

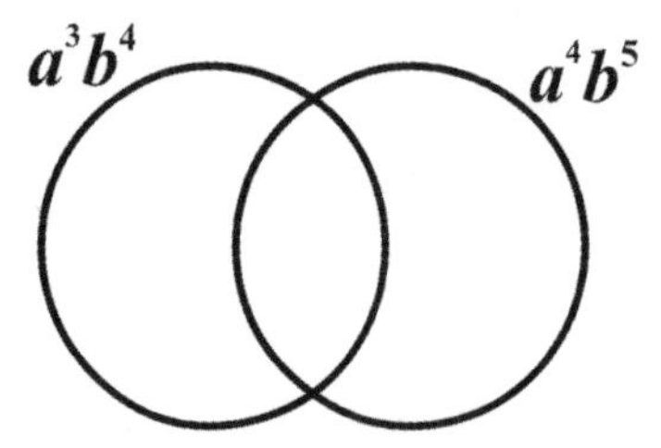


$c^2d; c^3d^4$

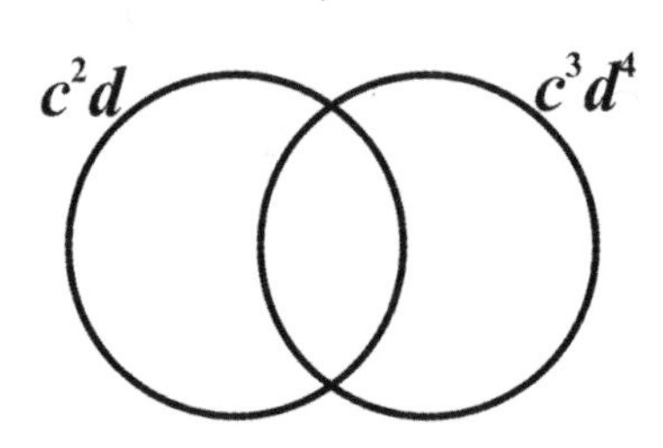

# Factor Lattices 1

Name: ______________________________ Date: ____________________

1. Draw a factor lattice to illustrate the product $x^5y^3$. Use your factor lattice to identify the least common multiple (LCM) for:

   a. $x^2$ and $y$ b. $x^3$ and $y^2$

2. Examine the factor lattice below and identify the LCM.

   a. $xy$ and $y^3$ b. $x$ and $x^2y^2$

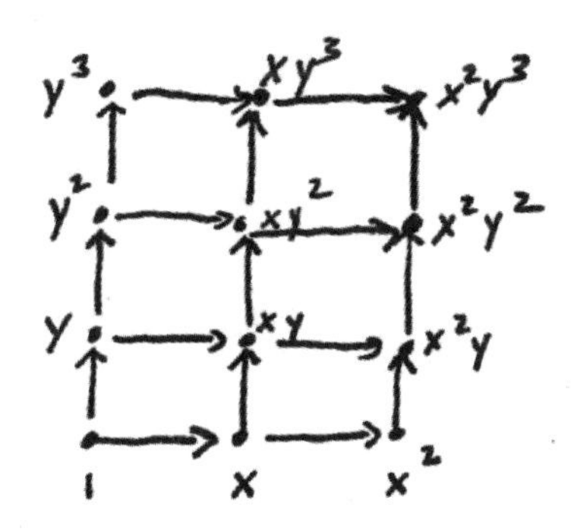


$x^2 y^3$

$x \rightarrow$ $y \uparrow$

3. Examine the same factor lattice above and identify the greatest common factor (GCF) for:

   a. $xy^3$ and $x^2y^3$ b. $x^2y^3$ and $y^2$

# Modeling Algebraic Expressions

Name: ______________________ Date: ______________________

The following expressions each represent the area of a parallelogram or rectangle. Use your Algeblocks and quadrant map or draw an area model to represent each expression. Sketch the resulting figure. For expressions in factored form, write the area. For expressions in expanded form, write the dimensions.

**1.** $b\,(b - 2)$

**2.** $-b(b - 2)$

**3.** $(b + 3)(b + 3)$

**4.** $b^2 + 5b + 6$

**5.** $b^2 - 5b - 6$

**6.** $b^2 - 6b + 8$

**7.** $(b - 2)(b - 2)$

**8.** $b^2 - 5b - 24$

**9.** $(b + 3)(b - 3)$

**10.** $b^2 - 16$

# Expanding Products 1

Name: ______________________________ Date: ____________________

Model each expression with Algeblocks and sketch your model. Write an equivalent expression in expanded form.

**1.** $(a + 2)(a + 3)$

**2.** $(a^2 + 2a + 1)(a + 1)$

**3.** $(a + 3)(a^2 + 5a + 6)$

**4.** Challenge: $(a + 3)^2(a + 2)^2$

# Expanding Products 2

Name: ______________________ Date: ______________

Model using algebra tiles or Algeblocks and draw a pictorial representation for the following. Solve each problem.

1. The area of a box is $a^2 + 4a + 3$ square inches. What might be the length and width of the box?

2. The following drawing illustrates the area of a plot of land that is $16x^2 + 48x + 20$ square feet. What might be the dimensions of the land?

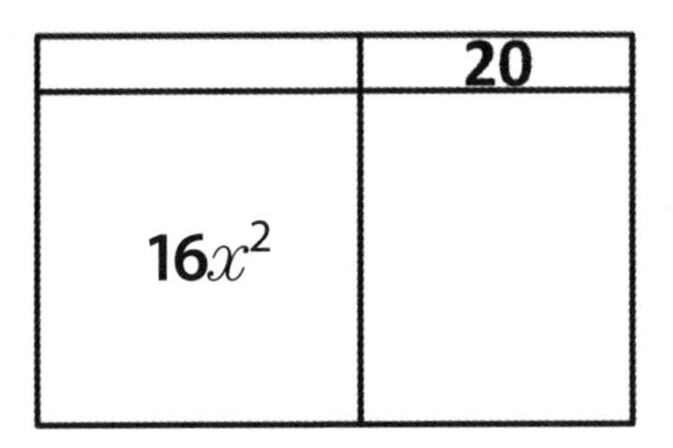


3. The volume of a container is $x^3 + 6x^2 + 11x + 6$ cubic feet. The height is $x + 1$ feet. What might be the length and width of the container?

# Don't Get FOILed Again!

Name: ______________________ Date: ______________



**1.** Ariel blocked off a portion of her yard in which to practice her gymnastics. After she made a square area, she decided to add 6 feet to both the length and the width of that area.

a. Create a model to represent the situation. Write an expression to match your model.

b. How much land did Ariel block off? (Be sure to include units of measure.)

**2.** Deion is helping his dad design a new, square walkway. They are adding an additional 3 feet to the length and width of the current square walkway. They are going to fill the new walkway with 6 inches of pea stone.

a. Create a model to represent the situation. What is the area of the new walkway? (Be sure to include units of measure.)

b. How much pea stone will there be when the walkway is complete? (Be sure to include units of measure.)

**3.** Here are the dimensions of a box. What is the volume of the box? Show your work.

height = $4x - 5$ feet

length = $3x + 1$ foot

width = $2x - 3$ feet

# Operating with Exponents

Name: ______________________ Date: ______________________

1. Explain why each of the following statements is true.

   a. $(2^3)(2^5) = 2^8$ b. $(5^7)(5^4) = 5^{11}$

2. Complete the following equation to show how you can find the exponent of the product when you multiply two powers with the same base. Justify your reasoning.

   $(a^m)(a^n) =$ __________

3. Explain why each of the following statements is true.

   a. $(2^3)(3^3) = 6^3$ b. $(5^7)(6^7) = 30^7$

4. Complete the following equation to show how you can find the exponent of the product when you multiply two powers with different bases. Your equivalent expression should have only one base. Justify your reasoning.

   $(a^m)(b^m) =$ __________

5. Explain why each of the following statements is true.

   a. $4^2 = (2^2)^2 = 2^4$ b. $9^2 = (3^2)^2 = 3^4$ c. $125^2 = (5^3)^2 = 5^6$

6. Write an expression having only one exponent for the following. Explain how you can find the base and the exponent when a power is raised to a power. Justify your reasoning.

   $(a^m)^n$

7. Write an equivalent expression that has only one prime base for each of the following.

   a. $\frac{1}{2}(2^n)$ b. $25(5^{-2n})$ c. $4^{n-1}$

# Powerful Integers

Name: ______________________________ Date: ____________________

1. Rewrite each expression using exponents. Then write how you would say that exponential expression in words. Evaluate the expression.

   a. $3 \times 3 \times 3$

   b. $(-5) \times (-5) \times (-5) \times (-5)$

   c. $b \times b \times b \times b \times b \times b \times b$

2. Rewrite each expression with a single base and a single exponent. Then write how you would say that exponential expression in words. Evaluate the expression.

   a. $(4^3)(3^3)$

   b. $(2^{-5})(5^{-5})$

   c. $(m^4)(n^4)$

   d. Write a verbal rule for multiplying powers with different bases and the same exponent.

3. Rewrite each expression with a single base and a single exponent. Then write how you would say that exponential expression in words. Evaluate the expression.

   a. $(5^3)(5^2)$

   b. $(3^{-7})(3^5)$

   c. $(-n)^4(-n)^8$

   d. Write a verbal rule for multiplying powers with the same base and different exponents.

**4.** Rewrite each expression with a single base and a single exponent. Then write how you would say that exponential expression in words. Evaluate the expression.

**a.** $\frac{3^5}{3^2}$

**b.** $\frac{5^3}{5^7}$

**c.** $\frac{6^{-4}}{6^7}$

**d.** $\frac{n^5}{n^3}$

**e.** Write a verbal rule for dividing powers with the same base and different exponents.

**5.** For each expression, write an equivalent expression using positive exponents.

**a.** $3^{-4}$

**b.** $\frac{1}{4^{-3}}$

**c.** $m^{-5}$

# Where Do I Go?

Name: ______________________________ Date: ____________________

Represent each situation on a number line and in an expression.

1. Tamara was three blocks east from her house when she received a call from Nicole. Nicole asked her to walk five blocks along the street her house is on to meet her at the yogurt shop. How far from Tamara's house (and in what direction) might the yogurt shop be?

2. Parker and Lucas left football practice at the same time. Parker had to walk seven blocks and Lucas had to walk thirteen blocks. How far apart might Peter and Lucas have ended up?

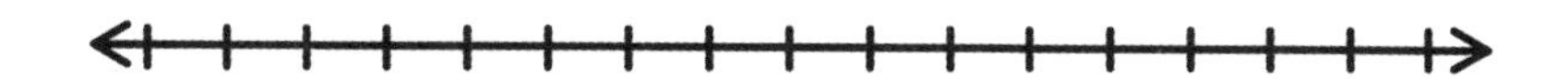

3. Kaylee and Sam are on the same soccer team. Kaylee started her practice exercises 5 yards in front of the east goal. Sam began her exercises 15 yards in front of the west goal. The girls met at the middle of the field. The field is 100 yards long. How far did each girl travel?

# Absolutely!

Name: ______________________________ Date: ______________________

Represent each situation on a number line.

1. Mackenzie and Janelle are best friends. When they leave school, Mackenzie walks 7 blocks east and Janelle walks 10 blocks west. What is the distance between them at the end of their walk?

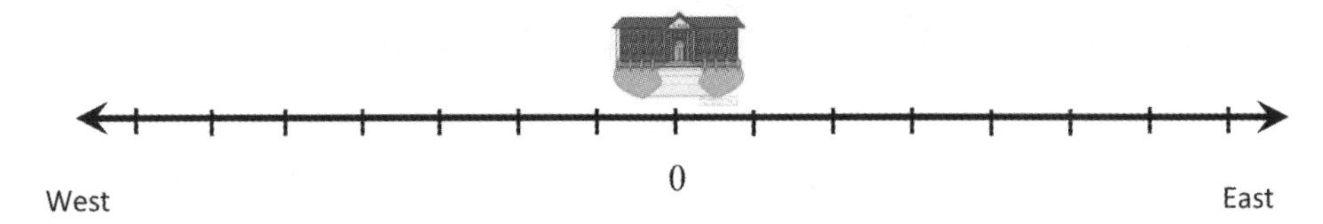


2. Edwin walks 20 blocks south from school to his gym. Jasmin walks 5 blocks north from school to babysit. Who finishes farther from school?

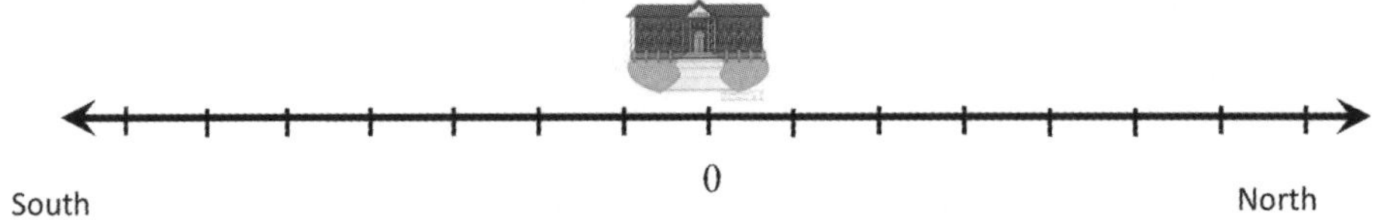


3. Jen-Min loves to go scuba diving. The deepest she has ever been is −60 feet. How far below sea level has she dived?

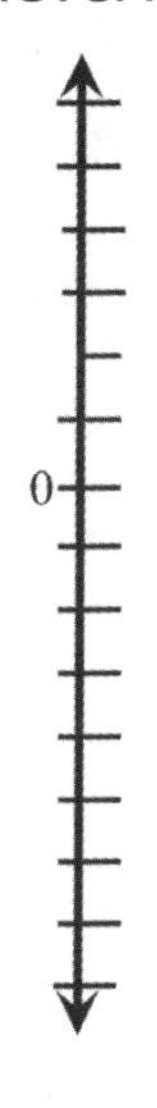

# Broken Calculator

Name: ______________________________ Date: ______________________

Sasha dropped her calculator when she got off the school bus, and it got banged up pretty badly. She gingerly picked it up and started doing calculations to see what the damage was. She was happy to find out it seemed to be working fine, despite all the scratches and dents. However, when she tried to compute a square root, she realized the square root key didn't work. Help Sasha approximate or find an exact value (if possible) for the following square roots, but pretend you're using her calculator, so you can't use your calculator to compute them. *In fact, try them without a calculator at all!*

**1.** $\sqrt{7}$

**2.** $\sqrt{29}$

**3.** $\sqrt{30.25}$

**4.** $\sqrt{51}$

# Babylonian Square Roots

Name: ______________________________ Date: ____________________

Model each expression with Algeblocks and sketch your model. Write an equivalent expression in expanded form.

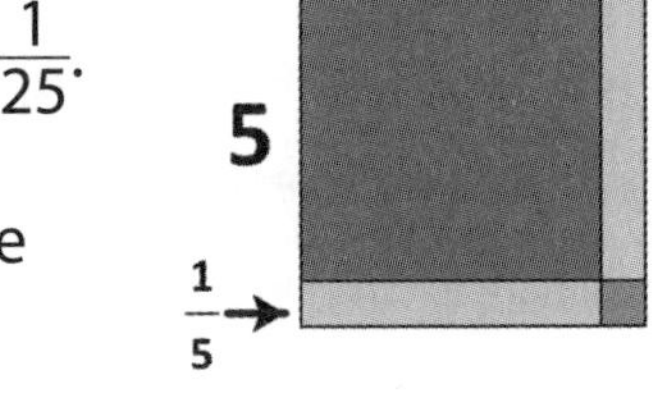


1. Explain how this figure shows that $(5\frac{1}{5})^2 = 27 + \frac{1}{25}$.

2. Label the side lengths in the figure below to use the Babylonian method to approximate $\sqrt{68}$.

right shape, wrong area

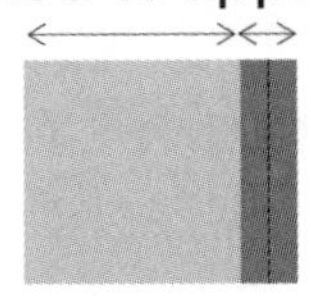
right area, wrong shape

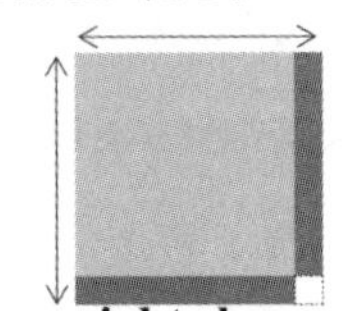
right shape, closer area

3. Label the side lengths in the figure below to use the Babylonian method to approximate $\sqrt{106}$.

right shape, wrong area

right area, wrong shape

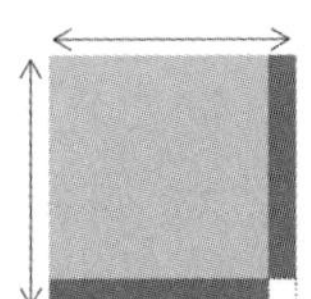
right shape, closer area

4. Use the Babylonian method to approximate $\sqrt{55}$.

right shape, wrong area

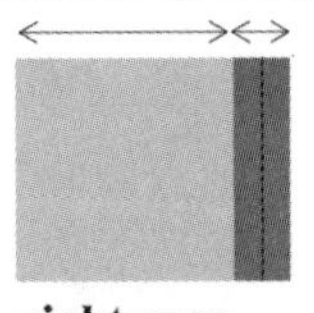
right area, wrong shape

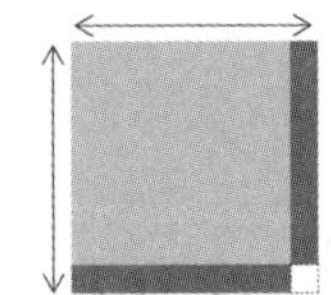
right shape, closer area

5. CHALLENGE: Use the figures below (and a written explanation) to show that the Babylonian approximation for $\sqrt{A^2 + b}$ is $A + \frac{b}{2A}$.

right shape, wrong area

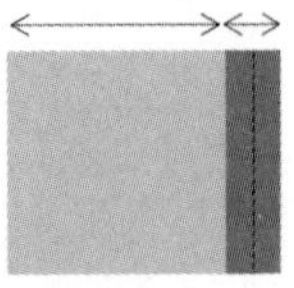
right area, wrong shape

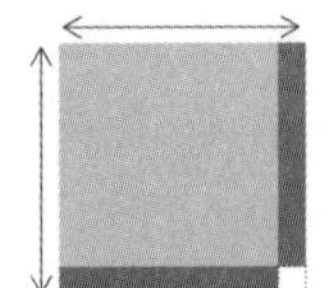
right shape, closer area

# Try This!

Name: ______________________ Date: ______________

A pretty good approximation for $\sqrt{68}$ is 8, because $8^2 = 64$. To find more decimals in the approximation of $\sqrt{68}$, we want to find $x$ so that $(8 + x)^2 = 68$.

Because $(8 + x)^2 = 64 + 2(8x) + x^2 = 64 + [2(8) + x](x)$, we know that $[2(8) + x](x) = 4$. We also know that $x$ is less than 1, so let's find its tenths digit.

Whatever this digit $d$ is, we know that $0.d \le x$, and we want $16.d \times d$ to be approximately 4, so we want to find the largest $d$ so that $16.d \times 0.d \le 4$. Experimenting with a few values, we have $16.1 \times 0.1 = 1.61$, $16.2 \times 0.2 = 3.24$, and $16.3 \times 0.3 = 4.89$, so 8.2 is our second approximation for $\sqrt{68}$.

Now $(8.2)^2 = 8^2 + 2(8)(0.2) + (0.2)^2 = 64 + 3.24 = 67.24$, which is closer, but still not equal to 68. Let's find the hundredths digit, $h$, of the approximation of $\sqrt{68}$.

If $(8.2h)^2 = 68$, then $68 = (8.2 + 0.0h)^2 = (8.2)^2 + 2(8.2)(0.0h) + (0.0h)^2$

$= 67.24 + (2(8.2) + 0.0h) \times (0.0h)$, so $0.76 = (2(8.2) + 0\ 0h) \times (0.0h) = 16.4h \times 0.0h$.

We wish we could find $h$ so that $16.4h \times 0.0h = 0.76$, but we'll settle for getting *close* to 0.76.

Because $16.44 \times 0.04 = 0.6576$ and $16.45 \times 0.05 = 0.8225$ (and 0.6576 is closer to 0.7600 than 0.8225), let's choose $h = 4$, so 8.24 is our new approximation for $\sqrt{68}$.

Checking by squaring 8.24, we get $(8.24)^2 = 67.24 + 0.6576 = 67.8976$, which is very close to 68.

Can you find the thousandths digit of the approximation? Fill in the missing details below:

$(8.24t)^2 = (8.24 + 0.00t)^2 = (8.24)^2 + 2(8.24)(0.00t) + (0.00t)^2$

$= 67.8976 + [2(8.24) + 0.00t](0.00t) = 67.8976 + (16.48t \times 0.00t)$

We want $16.48t \times 0.00t$ to be 0.1024:

$16.481 \times 0.001 = 0.016481$, $16.485 \times 0.005 = 0.082425$, $16.486 \times 0.006 =$ ______

$16.487 \times 0.007 =$ ______

We should choose $t =$ ______, so our next approximation is 8.24___.

Check by squaring your new approximation to be sure the result is still less than 68, but larger than 67.8976 (which equals $8.24^2$).

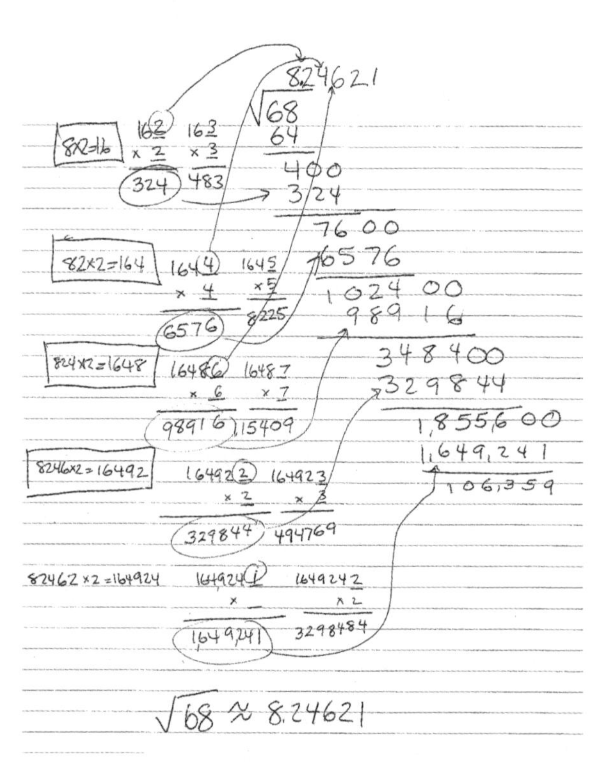


Can you see the connection between the work shown here and the work on the previous page? This square root approximation method can be "deciphered" by thinking about the work we did on the previous page. This method can be found in a number of books and websites.

# APPENDIX 2: Equations

# What Kind of Graph Am I?

Name: ______________________________ Date: ____________________

Cut out each figure and match the graph to its description. Glue into your notebook.

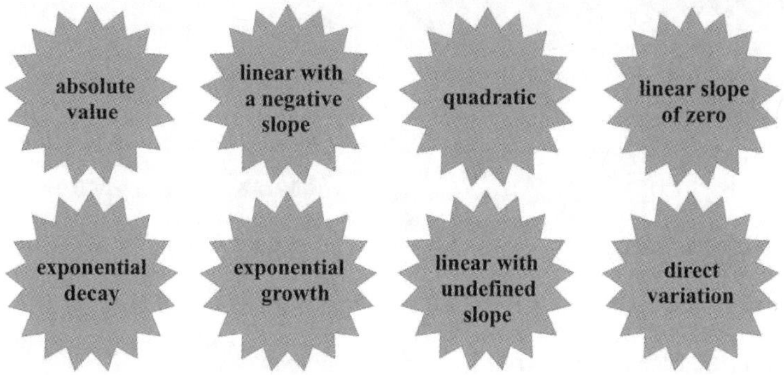

absolute value

linear with a negative slope

quadratic

linear slope of zero

exponential decay

exponential growth

linear with undefined slope

direct variation

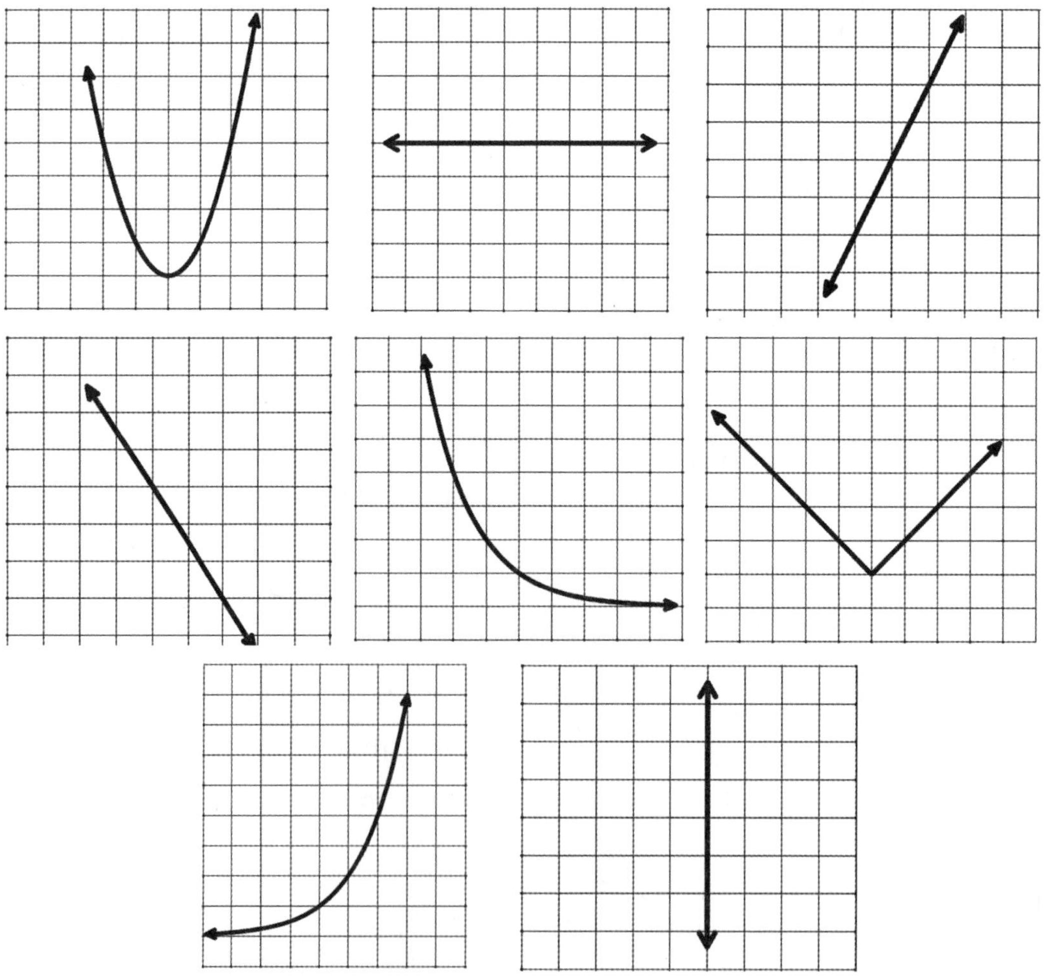

# Factor Lattices 2

Name: ______________________________ Date: ____________________

1. Draw a factor lattice to model the expression $xy^3$.

- Use your factor lattice to find the greatest common factor (GCF) for $xy^2$ and $xy^3$.

- Use your factor lattice to find the least common multiple (LCM) for $xy^2$ and $xy^3$.

2. Draw a factor lattice to model the expression $x^3y^4$.

- Use your factor lattice to find the GCF for $x^3y$ and $x^2y^3$.

- Use your factor lattice to find the LCM for $x^3y$ and $x^2y^3$.

# Balance Beams

Name: ______________________________ Date: ____________________

1. Find the missing values.

   a. $6 + 8 = \square + 3$

   b. $7 + \square = 2 + 25$

   c. $29 - 5 = 4 + \triangle$

   d. $\square \times 8 = 35 - 3$

   e. $16 - \square = 35 - 23$

   f. $4 + \square = 2 + \triangle$

2. Solve the following equations.

   a. $3a + 8 = 20$

   b. $6 - 4b = b + 6$

   c. $7 + 3c = c + 13$

   d. $2 - (3a + 4) = -6 + 4a$

   c. $-4 + 2(5b - 3) = 10b + 7$

# Can These Really Be True?

Name: ______________________________ Date: ______________________

Complete the following equations by inserting units of measure to make each statement true. For example: 1____ + 1____ = 11 could be 1 dime + 1 penny = 11 cents. There can be more than one correct answer.

1. 1________ + 1________ = 8________
2. 1________ + 1________ = 25________
3. 1________ + 1________ = 4________
4. 1________ + 1________ = 30________
5. 1________ + 1________ = 5________
6. 1________ + 1________ = 15________
7. Make up five equations of your own to challenge a friend.

# Mathematicians

### Carl Friedrich Gauss (1777–1855)

Born on April 30, 1777, in Brunswick, Germany, Gauss was a child prodigy in mathematics. The Duke of Brunswick was very impressed with Gauss's computing skills when he was only fourteen, so his stay at the Brunswick Collegium Carolinum, Hanover, was generously financed. Gauss was made the director of the Göttingen Observatory in 1807, as well as professor of mathematics at the same place. During his tenure, he spent much of his time establishing a new observatory. He also worked with Wilhelm Weber for almost six years making a primitive telegraph device that could send messages over a distance of 1,500 meters.

### Ada Lovelace (1815–1852)

Ada Lovelace's mother encouraged her to study science and mathematics. As an adult, Lovelace began to correspond with the inventor and mathematician Charles Babbage, who asked her to translate an Italian mathematician's memoir analyzing Babbage's Analytical Engine. (The Analytical Engine was programmed with punch cards and performed simple mathematical calculations. It is considered one of the first computers.) Lovelace went beyond completing a simple translation, however, and wrote her own set of notes about the machine. She even included a method for calculating a sequence of Bernoulli numbers; this is now acknowledged as the world's first computer program.

### Leonhard Euler (1707–1783)

Leonhard Euler was a pioneering Swiss mathematician and physicist. He made important discoveries in fields as diverse as infinitesimal calculus and graph theory. He also introduced much of the modern mathematical terminology and notation, particularly for mathematical analysis, such as the notion of a mathematical function. He is also renowned for his work in mechanics, fluid dynamics, optics, astronomy, and music theory.

Euler is considered to be the preeminent mathematician of the eighteenth century and one of the greatest mathematicians to have ever lived. He is also one of the most prolific mathematicians; his collected works fill sixty to eighty quarto volumes. He spent most of his adult life in St. Petersburg, Russia, and in Berlin, Prussia.

### Sophie Germain (1776–1831)

Unable to study at the École Polytechnique because she was female, Sophie Germain obtained lecture notes and submitted papers to faculty member Joseph Lagrange under a false name. When Lagrange learned she was a woman, he became her mentor. Germain soon began corresponding with other prominent mathematicians at the time. She became the first woman to win a prize from the French Academy of Sciences, for work on a theory of elasticity. Her attempted proof of Fermat's Last Theorem, though unsuccessful, was used as a foundation for work on the subject well into the twentieth century.

### Hypatia (ca. 350 or 370–415 or 416)

No one can know who the first female mathematician was, but Hypatia was certainly one of the earliest. She was the daughter of Theon, the last known member of the famed library of Alexandria, and followed his footsteps in the study of math and astronomy. She collaborated with her father on commentaries of classical mathematical works, translating them and incorporating explanatory notes, as well as creating commentaries of her own and teaching a succession of students from her home.

### Al-Khwarizmi (Ninth Century)

Muhammad Al-Khwarizmi was a brilliant Persian mathematician. He oversaw the translation of the major Greek and Indian mathematics and astronomy works into Arabic and produced original work that had a lasting influence on the advance of Islamic and European mathematics. The word *algorithm* is derived from the Latinization of his name, and the word *algebra* is derived from the Latinization of *al-jabr*, part of the title of his most famous book. In this book, he introduced the fundamental algebraic methods and techniques for solving equations.

Perhaps his most important contribution to mathematics was his strong advocacy of the Hindu numerical system, which Al-Khwarizmi recognized as having the power and efficiency needed to revolutionize Islamic and Western mathematics. The Hindu numerals 0–9, which have since become known as Hindu-Arabic numerals, were soon adopted by the entire Islamic world. Later, with translations of Al-Khwarizmi's work into Latin by Adelard of Bath and others in the twelfth century, and with the influence of Fibonacci's "Liber Abaci," they would be adopted throughout Europe as well.

# Let Me Count the Ways

Name: ____________________ Date: ____________________

1. Given the following rectangle, how many different rectangles are there altogether?

   a. Before solving the problem, predict how many different rectangles there are altogether.

   b. Write your solution and identify your problem-solving strategy.

   c. Write an expression to find the number of rectangles in a rectangle that is $n$ units long.

2. Given a 6-by-6 square, how many different squares are there altogether?

   a. Identify your problem-solving strategy.

   b. Write an expression to find the number of squares in a square that is $n \times n$ units large.

3. Write an expression for the following table.

| Input | 1 | 2 | 3 | 4 | 5 | 6 | $n$ |
|---|---|---|---|---|---|---|---|
| Output | 1 | 3 | 6 | 10 | 15 | 21 | |

   a. Using dots, draw a representation of the table data. Use the Input as the number of the figure.

   b. Describe what is happening to the figures you drew.

# Gauss Center Activity Cards

Name: ______________________________ Date: ______________________

**Counting Rectangles 1**

How many rectangles are embedded in a 2-by-3 rectangle, or a rectangle formed by two rows and three columns? Recall, you need to count all the rectangles in the entire shape.

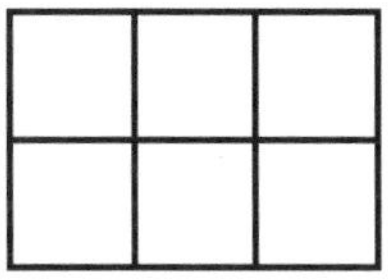

**Counting Rectangles 2**

How many rectangles are embedded in a 3-by-4 rectangle? Recall, you need to count all the rectangles in the entire shape.

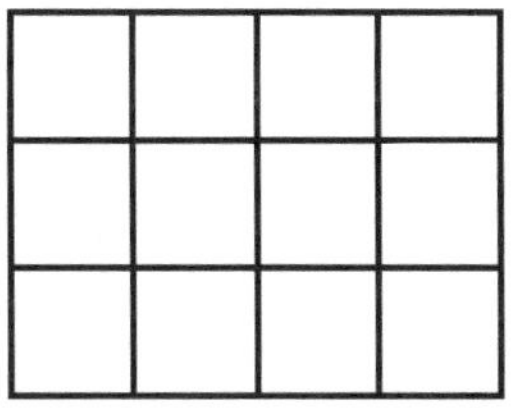

**Counting Rectangles 3**

How many rectangles are embedded in a 6-by-5 column rectangle? Recall, you need to count all the rectangles in the entire shape.

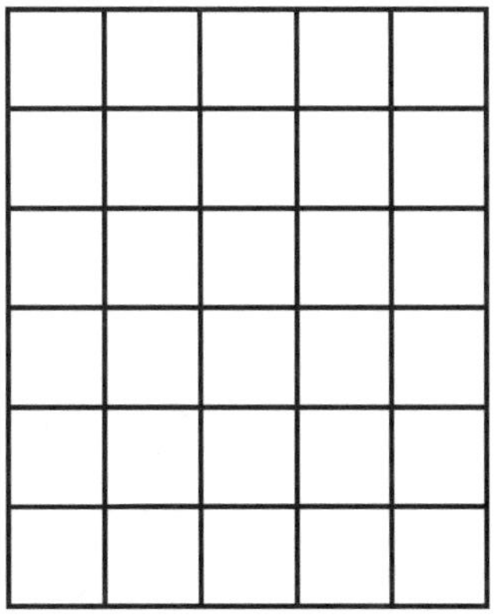

# Mobile Madness

Name: ______________________________ Date: ______________________

Find the values for each of the shapes. List or show the steps you use or explain your thinking.

1.

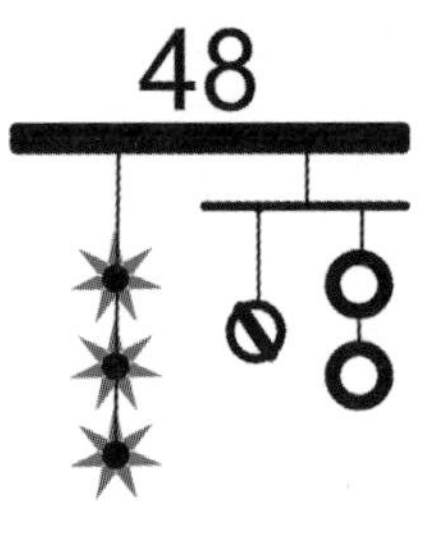


2.

72

♥= ___ ●= ___

3.

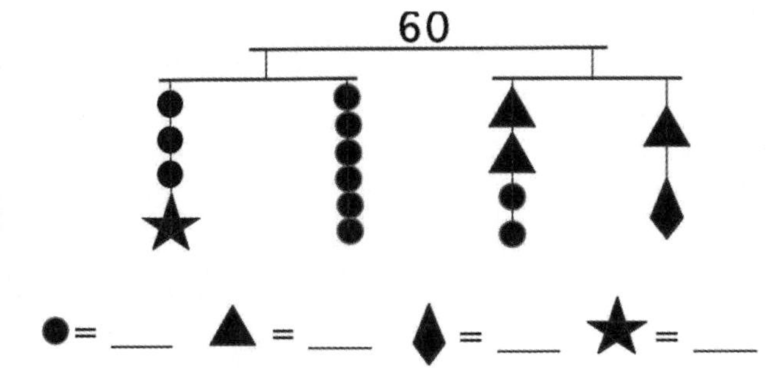


4.

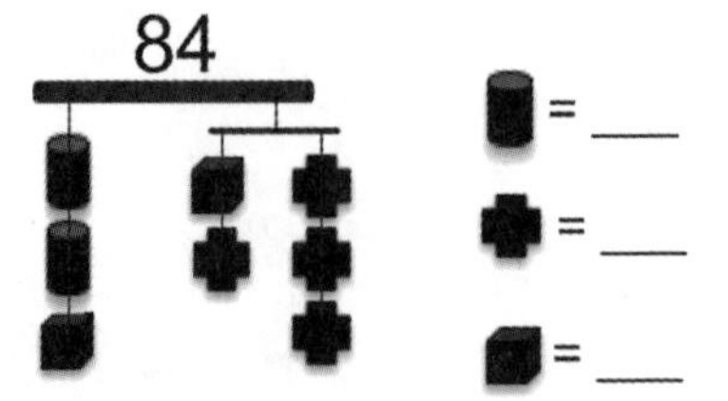

# Pattern Block Mobiles

Name: ______________________________ Date: ____________________

Examine the following figures.

**1.** Replace the **?** with one or more pattern blocks to balance the mobile. Can you find more than one way to do this?
If a triangle weighs 2 kilograms, how much does the mobile weigh (assuming that the bars and strings are weightless)?

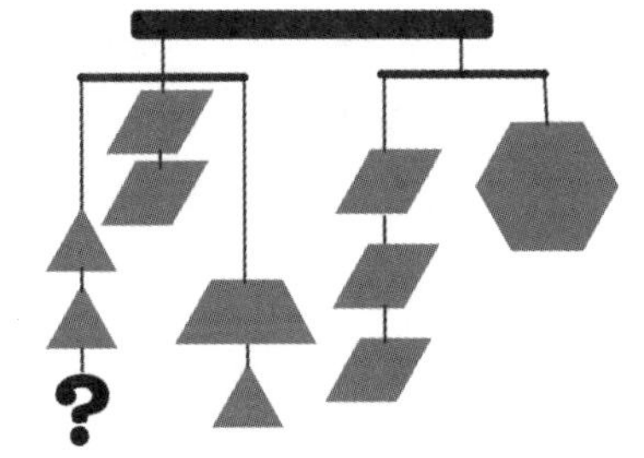


**2.** If you hung this mobile, would it balance? If not, what pattern block(s) could you add to balance it? Where?
If the entire mobile weighs 14 kilograms, how much does a triangle weigh?

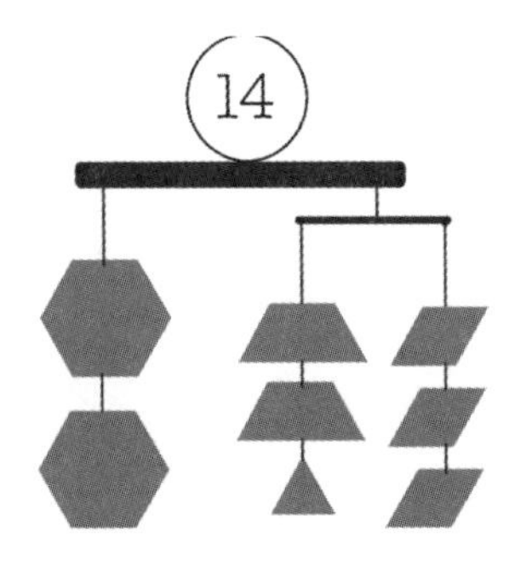


**3.** There are four extra pattern blocks for this partially constructed mobile. Place the extra blocks so that the mobile balances.

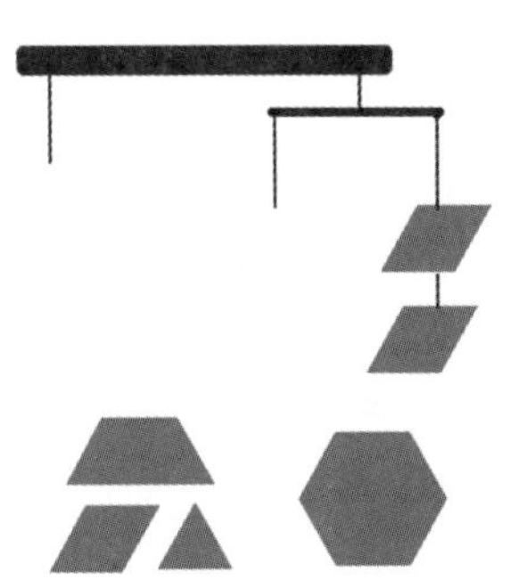

# The Knight's Move Challenge

Name: ______________________________ Date: ______________________

In chess, a knight has a distinct manner in which it can move. It may move two squares to the right and one square forward, two squares to the right and one square back, two squares to the left and one square forward, two squares to the left and one square back. The knight may jump over other players during a move. Your challenge is to move the knight so that it lands on each and every square of regular chess board in the *fewest moves* possible.

How many moves will it take to land on each square if the knight begins in the upper left square of the chessboard?

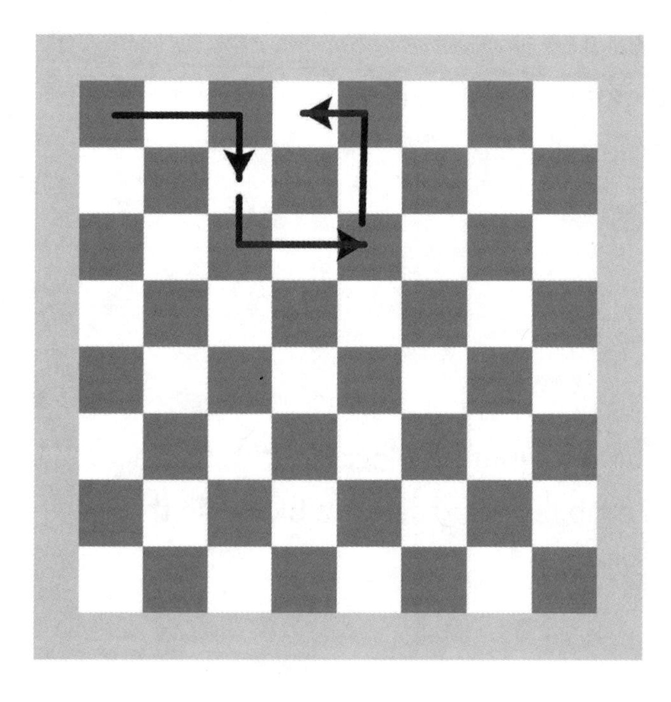

# Extra Chessboards

Name: ______________________________ Date: ______________________

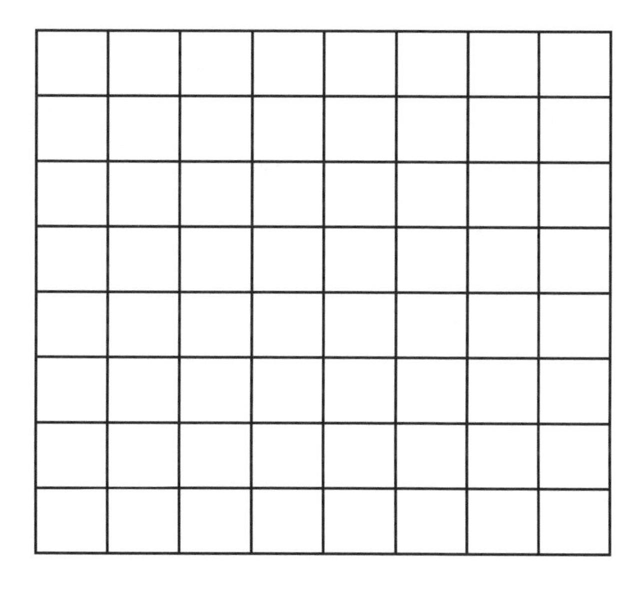

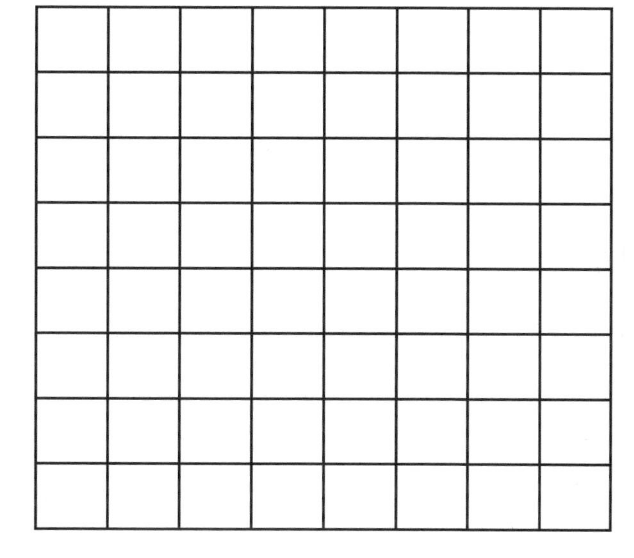

# Something Is Fishy

Name: ______________________________ Date: ______________________

Plot the ordered pairs on a Cartesian coordinate plane in the order they appear, from top to bottom. Connect points until you come to the word *STOP*. Start again with the next set of ordered pairs.

| START | (1, -13) | (-1, -16) | STOP | (2, 12) | (12, -4) |
|---|---|---|---|---|---|
| (-7, -11) | (2, -13) | (0, -19) | (-1, -14) | (2, 9) | (0, 15) |
| (-7, -13) | (4, -14) | (1, -20) | (0, -16) | (1, 6) | (-1, 12) |
| (-6, -15) | (6, -16) | (2, -20) | (1, -17) | (2, 3) | (-2, 10) |
| (-5, -16) | (8, -17) | (3, -19) | (2, -17) | (2, 7) | (-9, 3) |
| (-4, -16) | (9, -17) | (4, -16) | (3, -16) | (3, 9) | (-12, -1) |
| (-2, -15) | (10, -16) | STOP | (4, -14) | (2, 12) | (-13, -3) |
| (-1, -14) | (8, 5) | (-3, 17) | STOP | STOP | (-14, -10) |
| (-13, -15) | (6, 9) | (-2, 14) | (5, -8) | (-13, -5) | (13, -2) |
| (-11, -20) | (3, 14) | (-4, 12) | (4, -9) | (-16, -6) | (13, 1) |
| (-8, -23) | (2, 15) | (-5, 10) | (5, -10) | (-17, -8) | (14, -2) |
| (-4, -24) | (1, 17) | (-5, 7) | (6, -9) | (-17, -9) | (14, -4) |
| (3, -24) | (0, 20) | STOP | (5, –8) | (-16, -11) | (13, -6) |
| (6, -23) | (0, 22) | (-2, -8) | STOP | (-16, -10) | (15, -8) |
| (9, -21) | (1, 24) | (-3, -9) | (-13, -5) | (-15, -8) | (16, -10) |
| (11, -17) | (-2, 22) | (-2, -10) | (-15, -5) | (-13, -6) | (11, -6) |
| (12, -12) | (–3, 20) | (-1, -9) | (-16, 1) | STOP | END |
| (11, -5) | STOP | (-2, -8) | (-16, -3) | (11, -5) | |

# Cartesian Coordinate Plane

Name: ______________________________ Date: ______________________

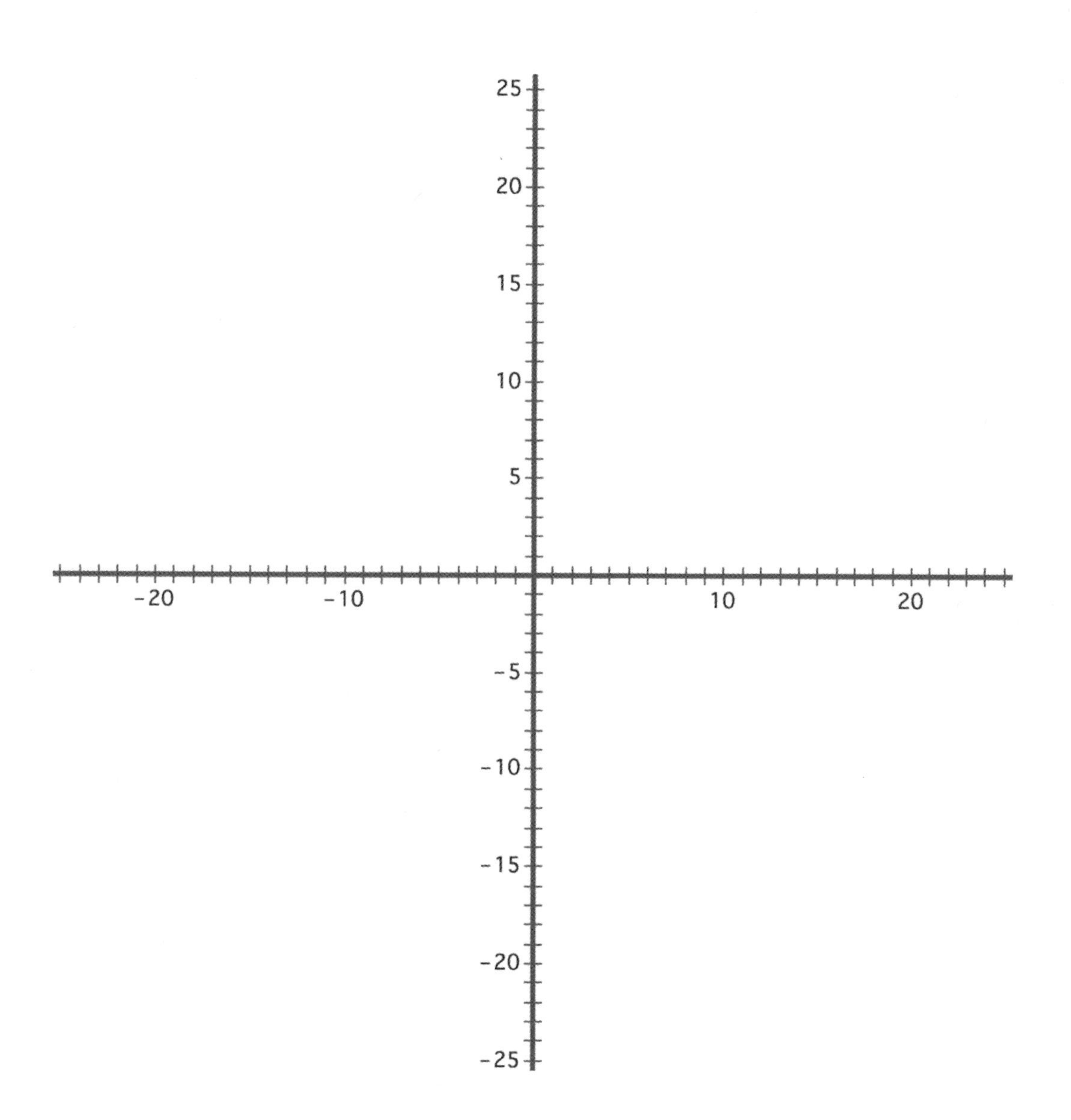

# Shape Equations

Name: ______________________ Date: ______________________

Determine the value for each symbol. Each symbol represents the same value whenever it occurs.

**1.**

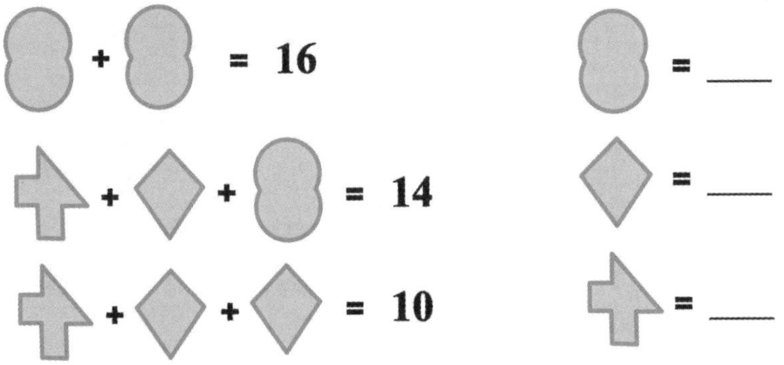

+ = 16 = ___

+ + = 14 = ___

+ + = 10 = ___

**2.**

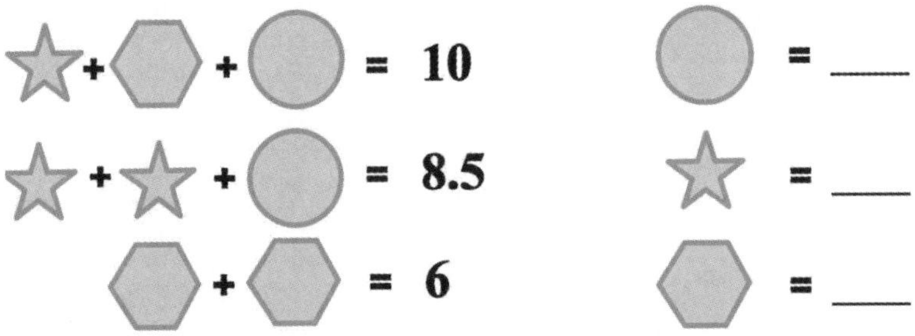

+ + = 10 = ___

+ + = 8.5 = ___

+ = 6 = ___

**3.**

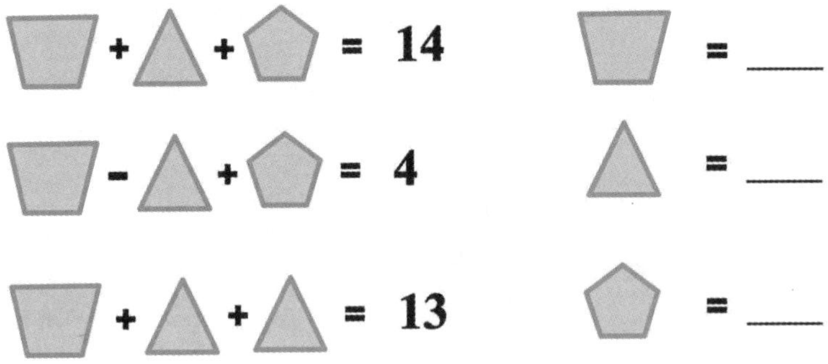

+ + = 14 = ___

- + = 4 = ___

+ + = 13 = ___

# Hypatia Center Activity Cards

Name: ______________________________ Date: ____________________

**Card A**

Complete the patterns. Assume the pattern continues to grow in the same way. Write a written rule or explain how you found the missing terms for each.

**a.** 1, 3, 6, 10, 15, _____, _____, 36, _____, 55

**b.** 1, 2, 3, 5, 8, _____, _____, _____, _____, 89

**c.** 1, 8, 27, 64, _____, _____, _____

**d.** 1, 3, 7, 15, 31, _____, _____, _____

**e.** 1, 4, 9, 1, 6, 2, 5, 3, 6, 4, 9, 6, 4, _____, _____, _____

**Card B**

Complete the patterns. Assume the pattern continues to grow in the same way. Write an expression or written rule for each.

**a.** 3, 7, 12, 18, 25, _____, _____, _____, _____

**b.** 0, 3, 0, 4, 1, 6, 3, _____, _____, _____, _____

**c.** 1, 3, 2, 4, 3, 5, 4, _____, _____, _____, _____

**d.** 256, 128, 64, 32, 16, 8, _____, _____, _____, _____

**Card C**

If the pattern continues, how many faces will be in the seventh, tenth, and *n*th row? Justify your response.

# Al-Khwarizmi Activity Center Cards

**Teacher:** Copy the cards on card stock. Cut out and mix them up. Students will match the situation with the number line with the inequality. Students must justify their responses.

Neelia has less than forty-five dollars to spend on books and magazines. She wants to buy a book for twenty-one dollars and spend the rest on magazines. Each magazine costs three dollars. At most, how many magazines can she buy?

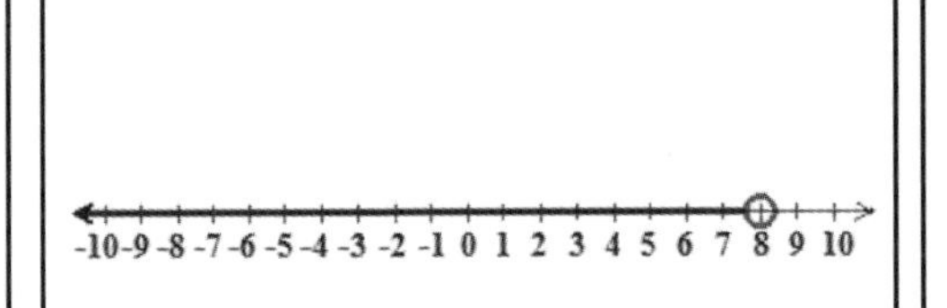


$3m + 21 < 45$

Analisa bought three identical packages of marbles and gave her friend twenty-one marbles. Analisa now has at least forty-five marbles. At least how many marbles can there be in each package?

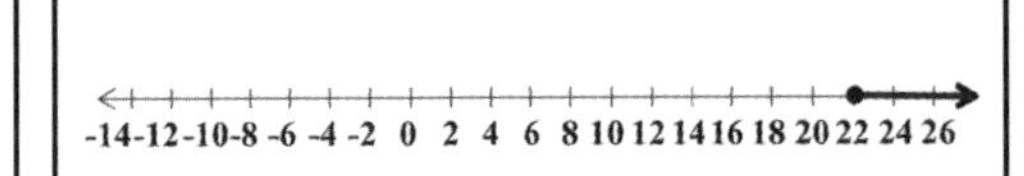


$3m - 21 \geq 45$

Zamira drove to the mall and back three times last week. Her brother drove the car a total of 21 miles last week. If together they drove no further than 45 miles, how far is one round trip to and from the mall?

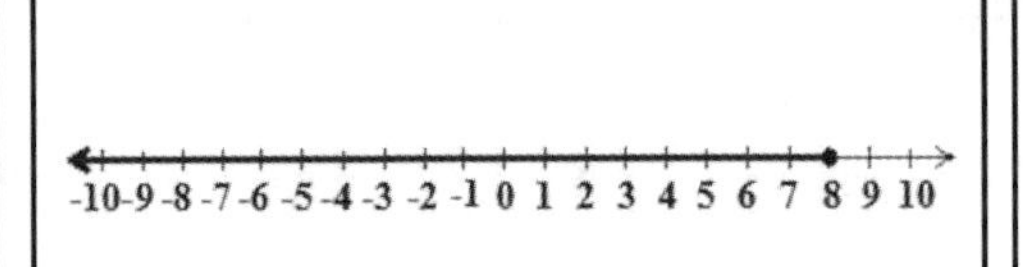


$3m + 21 \leq 45$

Johann's mother bought model airplane kits for three dollars each for Johann's birthday party. She also paid twenty-one dollars for party refreshments. If she spent more than forty-five dollars for the party, how many model kits could Johann's mother have purchased?

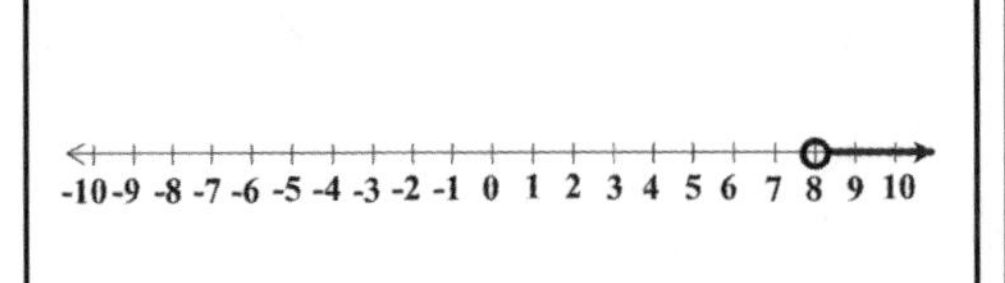


$3m + 21 > 45$

# Equations in All Forms

Name: ______________________________ Date: ____________________

1. Mariana sketched a line on graph paper for class. The points (2, 7) and (6, 19) lie on the line. Identify the slope as a ratio and write an equation for the line.

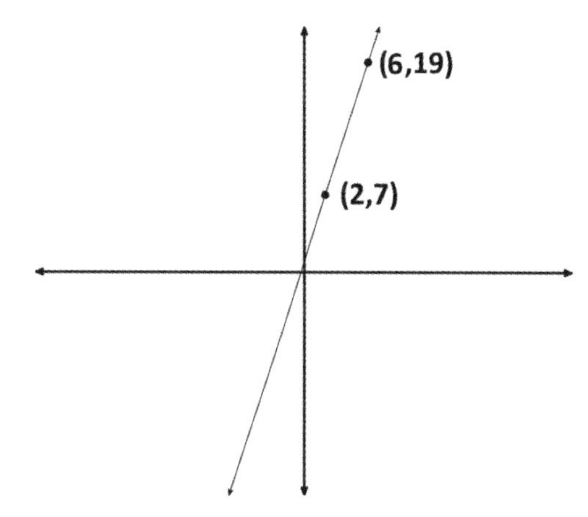


2. Given the following graph, identify the slope as a ratio and write an equation for the line given the point (4, 4).

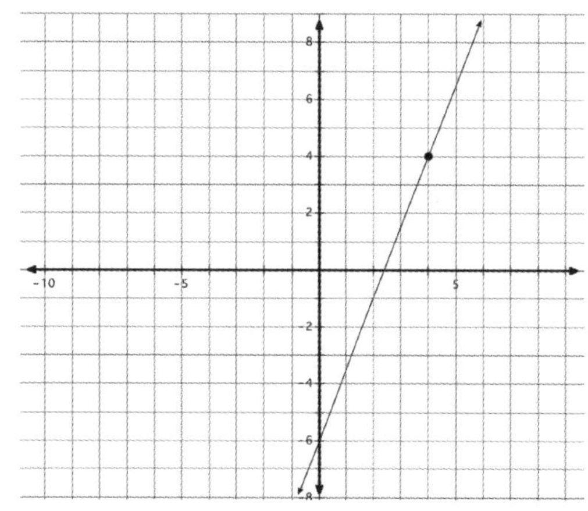

3. Nikolas plotted six points on the line satisfying the equation $2a + 3b = 12$. Complete the table by filling in the missing coordinates of the points he found.

| *a* | *b* |
|---|---|
| 3 | |
| | 8 |
| −3 | |
| | −2 |
| 0 | |
| | 0 |

# Graphs and Equations

Name: ______________________ Date: ______________________

1. Write an equation for the following ordered pairs. Identify the slope and graph the equation.

{(6, 0), (-4, 0), (3, 0), (-7, 0)}

2. Make a list of four ordered pairs that lie on the equation $2x + 3y = 12$. Identify the slope and graph the equation.

3. Write an equation for the following ordered pairs. Identify the slope and graph the equation.

{(0, -6), (0, 5), (0, 3), (0,-8)}

4. Identify the slope and graph the following equation.

$$3x + 15 = 6y$$

# Equation Puzzle

Name: ______________________________ Date: ____________________

Each shape has a different value. Each shape has the same value in all the equations. What is the value of equation D?

**A.** ⬠ + ◇ + ○ = 26

**B.** ⬠ + ○ + ○ = 23

**C.** ◇ + ○ = 15

**D.** ⬠ + ◇ = ?

# Linear Systems

Name: ______________________ Date: ______________

What strategies might you use to solve the following problems? Choose a different strategy to solve each of the following problems.

1. List your problem-solving strategies.

2. You and your friends are going to the mall to shop for spring and summer clothing. Your mom gives you $250 to spend. Capris cost $25 and dresses cost $50. You would like to purchase eight items. How many pairs of capris and how many dresses might you be able to buy?

3. Heather's mom is trying to decide between two plumber companies to fix her sink. The first company charges fifty dollars for a service call, plus an additional thirty-six dollars per hour for labor. The second company charges thirty-five dollars for a service call, plus an additional thirty-nine dollars per hour of labor. At how many hours will the two companies charge the same amount of money?

4. Two angles are supplementary. The measure of one angle is 30° smaller than twice the other. Find the measure of each angle.

# What Is Missing?

Name: ______________________________ Date: ____________________

1. Mariana sketched a line on graph paper for class. The points (2, 7) and (6, 19) lie on the line. Fill in the missing coordinates of additional points on the line. Graph the line and write an equation that best represents the graph.

   (4, ___) (0, ___) (1, ___)

   (___, 13) (___, 1) (___, 0)

2. Nikolas plotted six points on the line satisfying the equation $2x + 3y = 12$. Complete the table below by filling in the missing coordinates of the points he found. Identify the slope and the $y$-intercept.

| $x$ | $y$ |
|---|---|
| 3 | |
| | 8 |
| -3 | |
| | -2 |
| 0 | |
| | 0 |

3. Fill in the missing coordinates of additional points on the curve to satisfy the expression $x^2 - 6x + 8$. Graph the expression.

   (3, ___) (0, ___)

   (–2, ___) (2, ___)

# Graphic Organizer for Simultaneously!

Name: ______________________________ Date: ______________________

| | Number of Items | Value of Each Item | Value of Items |
|---|---|---|---|
| Item 1 | | | |
| Item 2 | | | |
| Item 3 | | | |
| Sum | | | |

| | Number of Items | Value of Each Item | Value of Items |
|---|---|---|---|
| Item 1 | | | |
| Item 2 | | | |
| Sum | | | |

| | Number of Items | Value of Each Item | Value of Items |
|---|---|---|---|
| Item 1 | | | |
| Item 2 | | | |
| Sum | | | |

# Simultaneously!

Name: ______________________________ Date: ____________________

Solve each of the situations below. Show all your work and justify your answers.

**1.** Coleman runs the Coolest Wheels Around Bike and Skate Shop. In one part of his shop, he has tricycles and bicycles. When he took inventory, he discovered he has 97 seats and 227 wheels.

a. How many tricycles does he have in stock?

b. How many bicycles does he have in stock?

**2.** The following month, Coleman's clerk counted the number of handlebars and wheels on the four-wheelers and the tricycles. He discovered he has 85 handlebars and 313 wheels.

a. How many tricycles did he have left in stock?

b. How many four-wheelers did he have in stock?

**3.** Across the street, the Bikes Unlimited shop advertised that it has the greatest selection of pedal vehicles. The store advertisement proclaims they have bicycles, tricycles, and unicycles. They brag that they have 777 wheels ready to roll, complete with 265 handlebars and 365 seats.

a. How many bicycles are at Bikes Unlimited?

b. How many tricycles are at Bikes Unlimited?

c. How many unicycles are at Bikes Unlimited?

# Hot Chocolate

Name: ______________________________ Date: ______________________

Avery and Connor are making hot chocolate for the cross-country ski team. They are trying to determine which of the following mixes they should use so the hot chocolate is really rich and chocolatey.

| *Parts Cocoa* | *Parts Milk* |
|---|---|
| 3 | 5 |
| 4 | 7 |
| 5 | 8 |
| 7 | 11 |

What strategies might you use to determine which of the mixtures Avery and Connor should use?

# Stephen and Cole



Stephen drove at a constant speed from town X to town Y at 9:00 a.m. yesterday. Half an hour later, Cole drove from town X to town Y at a constant speed that was 30 km/h faster than Stephen's. By 9:30 a.m., Stephen had already traveled 40 kilometers. Cole caught up with Stephen at town Y, arriving about the same time as Stephen.

a. At what speed was Stephen driving?

b. What was the distance between the two towns?

c. Graph the distance/time graphs for Stephen and Cole on the same axes.

# Which Region Is It?

Name: ______________________________ Date: ____________________

Sophia is in charge of ordering pizzas and cinnamon breadsticks for the Math Club Pi Day Celebration. A local pizzeria is offering a special deal for party orders: If you order a total of *at least* thirty-two items, the price will be \$10 per pizza and \$4 per order of cinnamon breadsticks. Sophia has a budget of \$200. How many pizzas and breadsticks might Sophia order?

1. What pair of inequalities represents the given information in the problem?

2. The graph below includes the graphs for the two equations $10p + 4b = 200$ and $p + b = 32$.

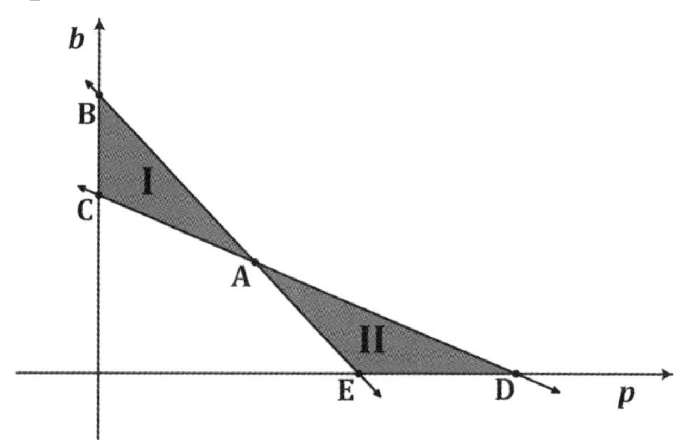


 a. Name the coordinates of the points labeled above.

 $A =$ $B =$ $C =$ $D =$ $E =$

 b. List all possible solutions in each region I and II.

3. Which region, I or II, best represents the pizza-cinnamon breadsticks combinations that Sophia can order? Justify your answer.

# Observation Protocol

| | Problem-Solving Strategy | | | Justifies Strategy and Solutions Using Accountable Talk | Accurate Computation |
|---|---|---|---|---|---|
| Student Name | Table | Graph | Equation | | |
| | | | | | |
| | | | | | |
| | | | | | |
| | | | | | |
| | | | | | |
| | | | | | |
| | | | | | |
| | | | | | |
| | | | | | |
| | | | | | |
| | | | | | |
| | | | | | |
| | | | | | |
| | | | | | |
| | | | | | |
| | | | | | |

# Solving Situations

Name: ______________________________ Date: ______________________

1. Ms. Wiley has $350 to spend on markers and individual whiteboards for her students. Markers cost $3.50 apiece, and whiteboards cost $6.25. How many markers and whiteboards might she purchase?

   a. Write an equation or inequality to represent this situation.

   b. Graph your equation or inequality.

   c. What is the optimal purchase?

2. A twenty-question quiz consists of true-or-false questions worth 3 points each and multiple-choice questions worth 11 points each. How many multiple-choice questions are on the test? (Assume the quiz is worth 100 points.)

   a. Write an equation or inequality for each variable to represent this situation.

   b. Graph your equation or inequality.

   c. How many true-or-false questions are on the test?

   d. How many multiple-choice questions are on the test?

3. When NASA began its space program, it advertised for astronauts. Male candidates had to be between 25 and 40 years of age and at least 5 foot, 4 inches tall but no taller than 5 foot, 11 inches.

   a. Write an equation or inequality that best represents the range of heights and ages for qualifying astronauts.

   b. Graph the equation or inequality that best represents the range of heights and ages for qualifying astronauts. Use graph paper and number your graph.

4. Maura makes five dollars an hour for babysitting and ten dollars an hour for tutoring. Because Maura is a full-time student, she cannot work more than 12 hours per week.

   a. Write two equations or inequalities that illustrate how many hours Maura must work at each job if she plans on earning less than ninety dollars per week.

   b. Graph the two equations or inequalities on one pair of axes.

# Backpacking

Name: ______________________________ Date: ______________________

Jordan went for a hike on Mt. Tom. When he got to the top of a cliff that is 121 meters high, he accidentally dropped his backpack off the edge. His friend Nicole was at the bottom of the cliff 43 meters from the base. She saw what was happening and ran to try and catch the backpack. She ran toward the base of the cliff as fast as she could at the exact moment the backpack was dropped. After 1 second, she was 35 meters away from the base.

- Draw a picture to illustrate what is happening.
- Model Nicole's distance from the base of the cliff as a linear equation of the time she began running. Define each variable.
- Use your model to predict how long Nicole took to reach the base of the cliff. Show your work.

# How Fast Did I Go?

Name: ______________________________ Date: ____________________

You are a highway patrol officer, seated on a motorcycle, on a curvy section of Highway 1. The posted speed limit is 45 miles per hour on this stretch of highway. You are monitoring traffic with a radar gun. The first exit is 3.6 miles up the road. Your radar picks up a speeding car traveling 68 mph. When you try to start your motorcycle to follow the car, it won't start. You try again and again, and soon you fear that you won't be able to catch the speeding car before it can turn off the highway. Finally, your motorcycle starts and you begin your pursuit 30 seconds after the speeding car has passed you on the roadside.

1. How fast do you need to go to catch up to the speeding car at the next exit, assuming that it continues to travel at 68 mph?
2. Illustrate, on a graph, the speed of the speeding car as well as your own motorcycle during this pursuit.
3. Is your own speed reasonable and safe?
4. Explain why this is or is not a good location at which to monitor traffic.

# Amanda's Subway Ride

Name: ______________________________ Date: ____________________

Amanda is taking the subway to meet some friends. She asks a fellow passenger how many stops there are until she gets to Park Street Station. The passenger tells her it is seven stops but neglects to tell her in which direction.

Write an equation and represent it on a number line.

# My Number Line Representation

**Teacher:** Copy (on card stock if possible) and cut out the cards. Instruct students to work in pairs and match the number line to the corresponding equation and solution card. Partners take turns making a match. The student with the most pairs is the winner.

| | | |
|---|---|---|
| $\lvert 3x - 6 \rvert = 12$ | -10 -9 -8 -7 -6 -5 -4 -3 -2 -1 0 1 2 3 4 5 6 7 8 9 10 | $x = 6$ or $x = -2$ |
| $\lvert 3x - 6 \rvert \le 12$ | -10 -9 -8 -7 -6 -5 -4 -3 -2 -1 0 1 2 3 4 5 6 7 8 9 10 | $-2 \le x \le 6$ |
| $\lvert -3x + 6 \rvert < 12$ | -10 -9 -8 -7 -6 -5 -4 -3 -2 -1 0 1 2 3 4 5 6 7 8 9 10 | $-2 < x < 6$ |
| $\lvert -3x + 6 \rvert \ge 12$ | -10 -9 -8 -7 -6 -5 -4 -3 -2 -1 0 1 2 3 4 5 6 7 8 9 10 | $x \le -2$ or $x \ge 6$ |
| $\lvert 3x + 6 \rvert \ge 12$ | -10 -9 -8 -7 -6 -5 -4 -3 -2 -1 0 1 2 3 4 5 6 7 8 9 10 | $x \le -6$ or $x \ge 2$ |

# Bottle Experiment Directions

**Teacher:** Provide triads of students with a 2- or 3-liter plastic bottle, a timer, duct tape, and a recording sheet, or have students bring notebooks. (The 3-liter bottle is less commonly available but results in a more distinct graph.)

Each group of students punches a hole, using an awl, into the side of the plastic bottle about 1 inch above the base. (You may also do this in advance.) The hole should not be very large, because the size will affect the rate at which the water flows. Once the hole has been made, students must seal the hole with duct tape.

Once the bottle is prepared, one student fills the bottle with water to the cap. (Because this can be messy, it is preferable to take students outside or provide each group with a bucket placed on a towel.) This student will control the flow of water, together with the student manning the duct tape. A third student will time how long the water is flowing. This student will call time every 10 seconds. When time is called, the hole is taped over, and the height of water left in the bottle is measured. This continues until the water level is below the hole.

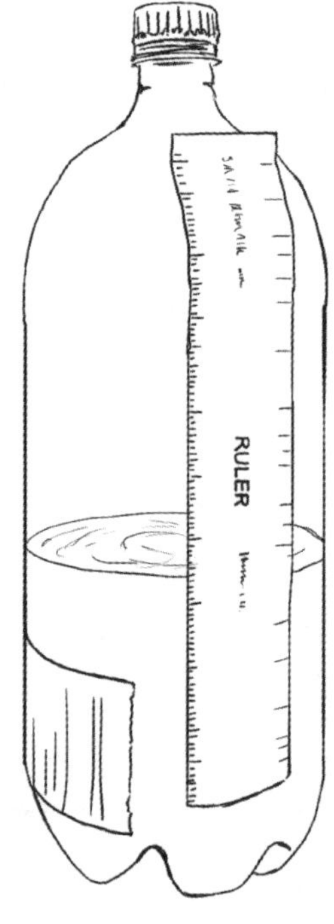


Students are advised to make a table to illustrate the amount of water over time and to graph their data. They should also write an equation for the line.

| **Water Height (centimeters)** | | | | | |
|---|---|---|---|---|---|
| **Time (seconds)** | | | | | |

# Am I Linear, Quadratic, Exponential, or Other?

Name: ______________________________ Date: ______________________

Examine the following tables and tell whether the data represent a linear, a quadratic, an exponential, or an absolute value relationship. Write an expression for the $n$th term.

**1.**

| Input | Output |
|---|---|
| 1 | 3 |
| 3 | 11 |
| 4 | 18 |
| 6 | 38 |
| $n$ | |

**2.**

| Input | Output |
|---|---|
| 2 | $\frac{1}{9}$ |
| -3 | $\frac{1}{-27}$ |
| 4 | $\frac{1}{81}$ |
| -5 | $\frac{1}{-243}$ |
| $n$ | |

**3.**

| Input | Output |
|---|---|
| 2 | 2 |
| 5 | 11 |
| 8 | 20 |
| 11 | 29 |
| $n$ | |

**4.**

| Input | Output |
|---|---|
| 8 | 8 |
| (-5) | 5 |
| 14 | 14 |
| (-7) | 7 |
| $n$ | |

# Delilah and the Rooftop

Name: ______________________________ Date: ____________________

Delilah (who is 6 feet tall) throws a ball off the roof of her 45-foot-high apartment building, so that it follows a path described by the equation: $h = -16t^2 + 128t + 51$, where $h$ is the height of the ball above the ground $t$ seconds after it leaves Delilah's hand.

1. How high above the roof will the ball be 2 seconds after Delilah releases it?

2. How many seconds after she releases the ball will it be at the same height as in question 1?

3. How high above the roof will the ball be when it reaches its highest point?

4. How much time will it take for the ball to fall to the ground?

5. How long will it take for the ball to reach a point that is 200 feet above the ground?

6. The ball is 207 feet above the ground after 1.5 seconds and also after _____ seconds.

7. Provide a complete, accurate, and appropriately labeled graph.

# Will She Catch It?

Name: ______________________________ Date: ____________________

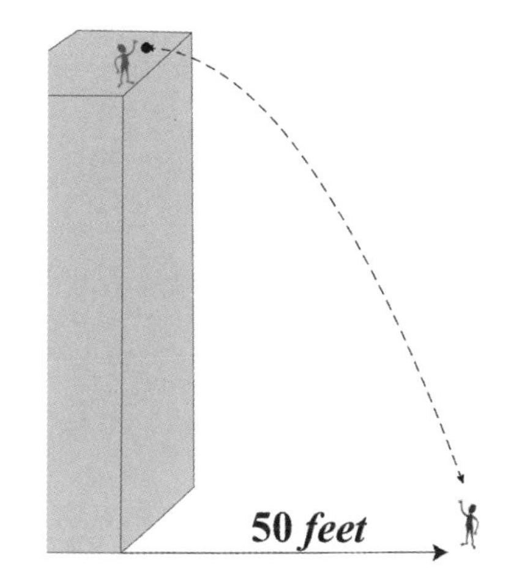


Sophia is standing at the top of her apartment building, which is across the street from a tennis court. She has a tennis ball and wants to try to throw it to her friend, Steph, who is standing on the tennis court, exactly 50 feet from the apartment building. Sophia has been doing some research and some calculating and has found that if she throws the ball horizontally, the ball's height in feet will be related to its horizontal position, $x$ (also measured in feet), by the function $h(x) = -Ax^2 + 105$, where $A$ is a number based on the speed the ball is thrown.

1. To get the ball to Steph, when the horizontal position of the ball is 50 feet from the building, the height of the ball needs to be 5 feet.

   a. Express "the height of the ball is 5 feet when its horizontal position is 50 feet" as an algebraic equation using the function $h(x)$.

   b. Solve for $A$ so that the equation you found in part a is true.

2. Sophia's research showed that $A = \frac{16}{v^2}$, where $v$ is the velocity at which the ball is thrown (in feet per second). Solve for $v$ so that $A$ is the value you found in part b above.
3. Your answer in item 2 is the velocity Sophia needs to throw the ball to get it to Steph. The unit is feet per second. Convert this velocity into units of miles per hour to see if this is a reasonable speed. (Most people throw between 20 and 50 miles per hour, though some baseball pitchers throw as fast as 90 and 100 miles per hour!)

APPENDIX 2: EQUATIONS

# Modeling Motion

Name: ______________________ Date: ______________

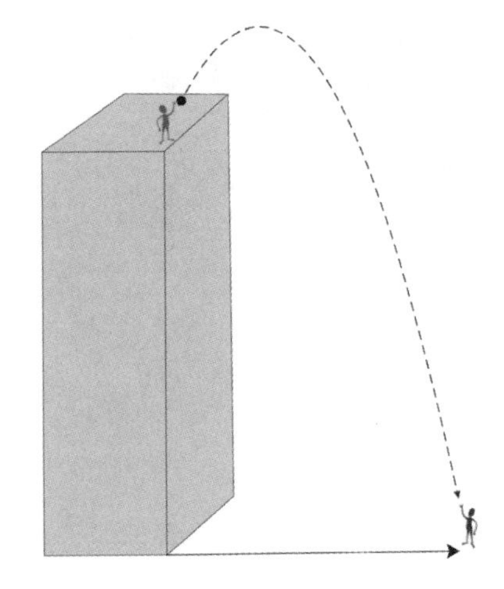

Things get a little more complicated when the ball is not thrown horizontally. Sophia determined that the height function in this case is $h(x) = -Bx^2 + Cx + 105$, where $B$ and $C$ both depend on the speed the ball is thrown *and* the angle at which it is thrown (with respect to horizontal). Sophia decided to attempt to throw the ball at an angle of 45° above the horizontal, resulting in the function $h(x) = -Bx^2 + x + 105$.



**1.** Make a guess: Do you think Sophia will have to throw the ball faster or slower (compared to when she threw the ball horizontally) to get the ball to Steph?

**2.** Steph will be able to catch the ball if the height of the ball is 5 feet when $x$ is 50 feet. Find $B$ so that this will be true.

**3.** Sophia determined that $B = \frac{32}{v^2}$, where $v$ is the velocity at which the ball is thrown. Solve for the value of $v$ that makes the equation true when $B$ is the value you found in item 1. The units will be feet per second.

4. Convert your answer to item 3 into miles per hour and compare this velocity to the solution to item 3 on the previous page. Is it still a reasonable speed to throw? Justify your response.

# APPENDIX 3: Functions

# Working with Functions, Part 1

Name: ______________________________ Date: ______________________

For each problem on this page, $f(x) = 3x - 4$.

1. Compute $f(1)$.

2. Compute $f(2)$.

3. Compute $f(3)$.

4. Compute $f(4)$.

5. Compute $f(5)$.

6. Does $f(2) + f(3)$ equal $f(5)$?

7. Does $f(3) - f(2)$ equal $f(3 - 2)$?

8. Compute and simplify $f(a + b)$.

9. Explain how your answer in item 8 shows that $f(a + b)$ does not always equal $f(a) + f(b)$.

10. Does $f(a) + f(b)$ ever equal $f(a + b)$?

11. Does $f(-a)$ ever equal $-f(a)$?

12. Does $f(-a)$ ever equal $f(a)$?

# Working with Functions, Part 2

Name: ______________________ Date: ______________

For each problem on this page, $g(x) = x^2 - 4x + 2$.

**1.** Compute the following outputs of the function $g(x)$:

a. $g(2) =$

b. $g(5) =$

c. $g(-2) =$

d. $g(3) =$

e. $g(7) =$

f. $g(-3) =$

**2.** Explain how you know that $g(-a)$ does not always equal $-g(a)$.

**3.** Does $g(-a)$ ever equal $g(a)$?

**4.** Explain how you know that $g(a + b)$ does not always equal $g(a) + g(b)$.

**5.** Does $g(a + b)$ ever equal $g(a) + g(b)$? Show how you know.

**6.** Does $g(-a)$ ever equal $-g(a)$? Show how you know.

**7.** Does $g(-a)$ ever equal $g(a)$? Show how you know.

**8.** Show that $g(a - b) = a^2 + b^2 - 2ab - 4a + 4b + 2$.

**9.** Show that $g(a) - g(b) = a^2 - b^2 - 4a + 4b$.

# Am I a Function?

Name: ______________________________ Date: ______________________

Add one example that is a function and one example that is not a function in the blank squares. Then cut out each of the squares and sort them into two columns: Functions and Not Functions.

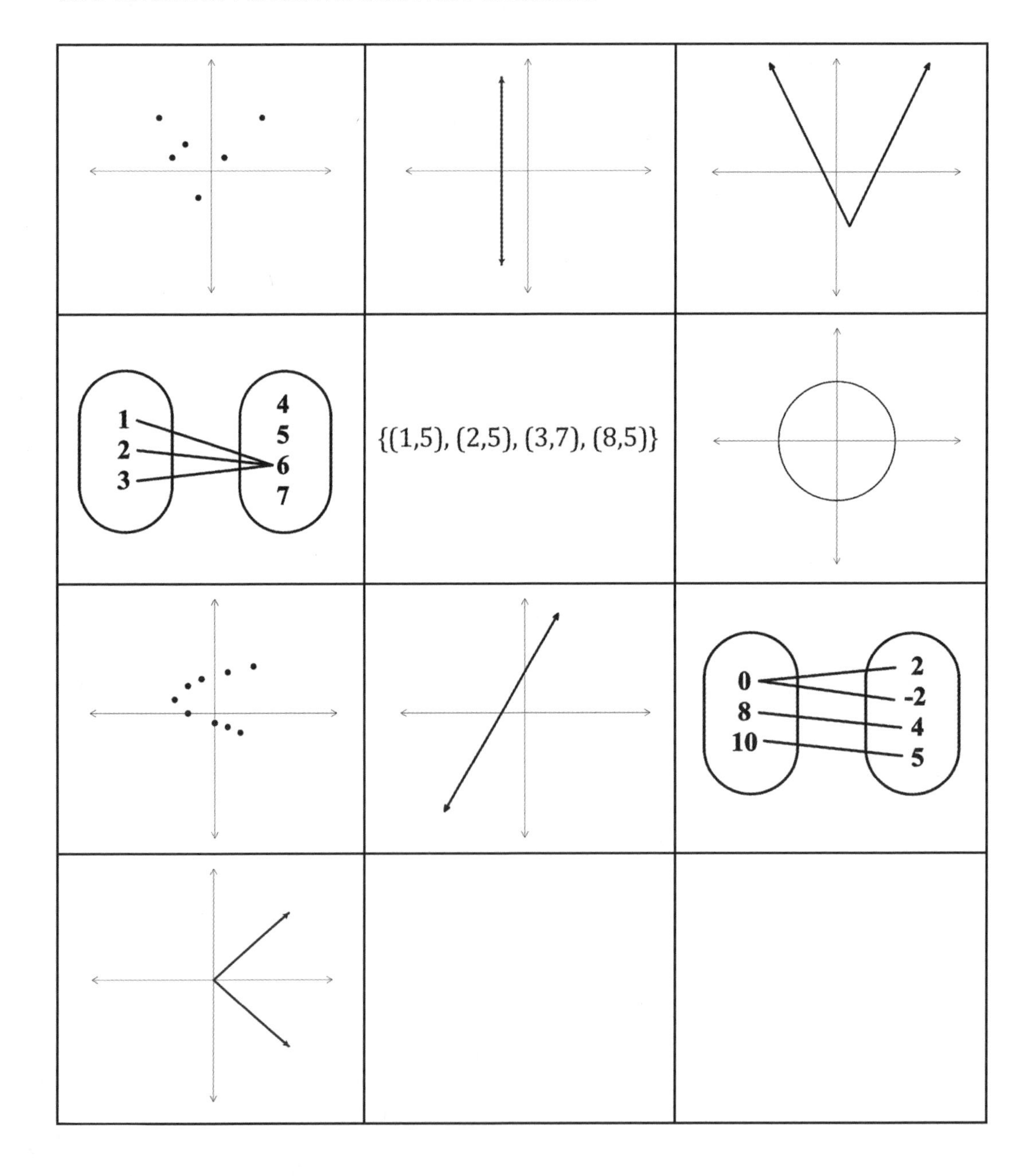

# Double-Wides

Name: ______________________ Date: ______________

The first four "double-wide" numbers—two, eight, eighteen, and thirty-two—are described visually in the figures.

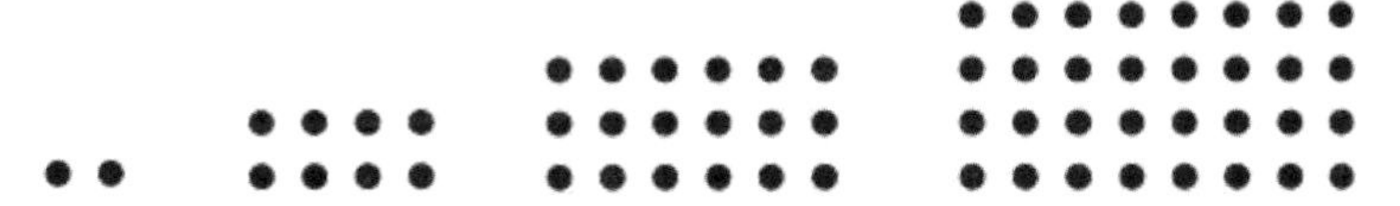

1. Can you see why they're called the double-wide numbers? Write a possible reason.
2. The function $D(n)$ represents the $n$th double-wide number whenever $n$ is a counting number. That is, $D(1) = 2$, $D(2) = 8$, $D(3) = 18$, and $D(4) = 32$. Compute the following double-wide numbers and explain your reasoning:
   a. $D(5)$
   b. $D(10)$
   c. $D(100)$
3. Find an algebraic expression that represents $D(n)$ and explain your reasoning.
4. Use the expression for $D(n)$ you found in item 3 to compute $D(2.5)$ and $D(-3.7)$. Explain why $D(2.5)$ and $D(-3.7)$ are *not* double-wide numbers.
5. Is 128 a double-wide number? Can you find *two* numbers, $x$ and $y$, so that $D(x)$ and $D(y)$ both equal 128? Explain your reasoning.
6. Can you find a number $x$ so that $D(x) = 75$? Is 75 a double-wide number? Explain your reasoning.
7. What is your prediction for the first four "triple-wide" numbers? Explain your reasoning.

# Square Numbers

Name: ______________________________ Date: ____________________

1. The first four square numbers are one, four, nine, and sixteen, as illustrated in the figure. Find the fifth and fifteenth square numbers.

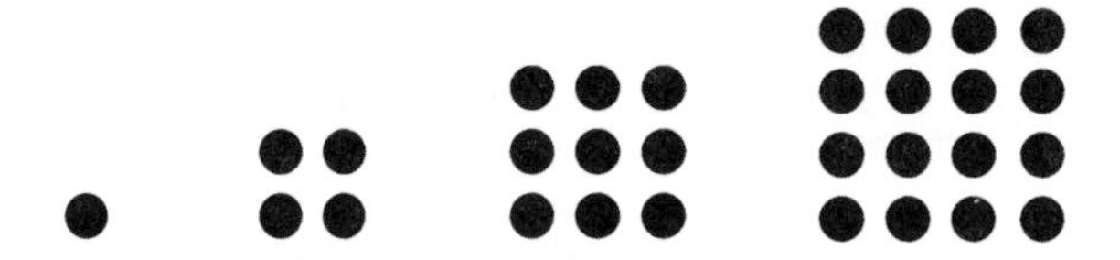

2. We will call $S(x)$ the *square number* function. That is, $S(1) = 1$, $S(2) = 4$, $S(3) = 9$, and $S(4) = 16$. Find an algebraic expression for the $n$th square number $S(n)$.

3. The square number function can be extended to inputs that aren't counting numbers. Using the expression for $S(n)$ you found in item 2, compute $S(1.5)$, $S(-3)$, and $S(\sqrt{7})$.

4. Even though 30.25 is *not* a square number, find a number $x$ so that $S(x) = 30.25$. The number you find won't be a counting number.

5. Find another number $x$ (besides the one you found in item 4) so that $S(x) = 30.25$.

6. Does every output for $S(x)$ have two inputs? Explain your reasoning.

# Translate This

Name: ______________________ Date: ______________

Ibrahim has designed an experiment for the National Oceanic and Atmospheric Administration to measure temperature and wind speed (and direction) at different altitudes. His measuring instruments have been attached to a weather balloon that was launched in the middle of the Atlantic Ocean. The mechanism holding the instruments to the balloon was designed to let the instruments loose when the height of the balloon is exactly 90,000 feet (about 17 miles) above sea level.

The function $f(x)$ represents the height (in feet) of the instruments exactly $x$ seconds after the instruments were released. Therefore, the values of $x$ can be any number of seconds between 0 and the number of seconds it takes until the instruments splash into the ocean.

1. Translate the following statements into English sentences or phrases describing the instruments. If possible, determine whether the statements are true, and explain how you know or why it's impossible to know.

   a. $f(0) = 90{,}000$

   b. $f(7) > f(15)$

   c. $f(15) = 95{,}000$

   d. $f(15) = 0$

   e. $f(90{,}000) = 0$

   f. $f(14) = f(15)$

2. Express the following quantities using function notation:

   a. The height of the instruments 1 minute after they were released from the balloon.

   b. The average velocity of the instruments from the time the instruments were dropped until the instruments had been falling for 5 seconds.

3. Ibrahim had read that $f(x) = 90{,}000 - 16x^2$ is a reasonable formula* for the height function. Use this formula and the functions you wrote in 2a and 2b to find the following quantities:

   a. The height of the instruments after 15 seconds

   b. The height of the instruments after 1 minute

   c. The average velocity of the instruments in the first 5 seconds after they were released

   d. The time it takes for the instruments to splash into the ocean

   *Note: When an object falls from such a distance, the air resists, or pushes against, the object and slows it down. This formula does not account for that air resistance.

# Translate That

Name: ______________________________ Date: ____________________

The cost of renting a car for 5 days from A1 Auto Rental depends on the number of miles you drive the car. The function $f(x)$ represents the cost (in dollars) of renting a car for 5 days and driving $x$ miles:

$$f(x) = \begin{cases} 200 & \text{if } x \leq 180 \\ 200 + .20(x - 180) & \text{if } x > 180 \end{cases}$$

**1.** How much will it cost to rent the car if you drive 150 miles? How much will it cost to rent the car if you drive 200 miles?

**a.** Which of these statements is correct? Explain your reasoning.

**a.** $f(180) = 200$ **b.** $f(200) = 180$

**2.** Fill in the blanks to complete the verbal description of the cost to rent the car.

*The cost is* $__________ *as long as you drive no more than* __________ *miles. For every mile beyond* __________ *miles that you drive, you will be charged an additional* __________ *cents.*

**3.** Rachel had to pay $250. How many miles did she drive?

**4.** Xenia said, "The number of dollars I had to pay to rent the car was the same as the number of miles I drove the car."

**a.** Express Xenia's observation as an equation using function notation, letting $m$ equal the number of miles she drove the car.

**b.** How many miles did Xenia drive the car?

# How Cold Is It?

Name: ______________________________ Date: ____________________

1. Sean was visiting family in Canada in October and had heard that it might get cold while he was there. One morning, he was listening to the radio during the weather forecast. The station's meteorologist said that the current temperature was 15° Celsius. Fortunately, Sean remembered that the conversion function from Celsius to Fahrenheit is $F(C) = 1.8C + 32$. What did Sean determine the current Fahrenheit temperature was? Show your work.

2. Sean's cousin, Yvette, said she had heard that the function $T(C) = 2C + 30$ is a good approximation to the Celsius-to-Fahrenheit conversion function.

   a. Compare the functions $F$ and $T$ at the following Celsius values: 5°, 10°, 15°, and 20°.

   b. Graph the functions $F$ and $T$ to check your results from part a.

   c. Determine whether there are any Celsius values for which $T(C) = F(C)$.

   d. Find all Celsius values for which $|T(C) - F(C)| \le 5$.

# Absolute Translations

Name: ______________________________ Date: ____________________

For each of the following sentences, determine the number (or numbers) that make the sentence true. Then rewrite the sentence using mathematical symbols and variables.

1. Five times a number equals thirty-five.

2. The square of a number is sixteen.

3. The difference between a number and five is three.

4. The difference between five and a number is three.

# Mix and Match

Name: ______________________________ Date: ____________________

Match each equation to the appropriate situation and graph.

a. $f(x) = 75x - 15$ b. $f(x) = 15x - 75$

c. $f(x) = 15|x - 75|$ d. $f(x) = 75|x - 15|$

**A.** The meteorologist at WMATH has a deal with the station manager. She has to pay fifteen dollars per degree for the difference between her forecasted high temperature and the actual high temperature. Her forecasted high temperature for today was 75°F. How much will she have to pay if the actual high temperature was $x$ degrees?

**B.** The school math club is having a family banquet and is expecting fifteen families to attend. The caterer has offered a special deal as long as exactly fifteen members attend, but the club will be fined if either fewer or more than fifteen families attend. (They do this to make up for the last minute changes in food or staff.) The club will have to pay an additional seventy-five dollars for each family *different* from fifteen. How much more will the club have to pay if $x$ families attend the banquet?

**C.** Marco is having a birthday party, and his uncle gave every party attendee seventy-five marbles each. After playing with the marbles for over an hour, fifteen marbles had been lost. If $x$ people attended Marco's party, how many marbles remained? (Assume that many of Marco's friends attended.)

**D.** Jorge bought the food and drinks for his nephew's birthday party, which cost fifteen dollars for each party attendee. Because Jorge was a loyal customer of The Leaning Tower of Pizza, he got a seventy-five-dollar discount off his total. If $x$ is the number of party attendees, how much did Jorge have to pay?

I.

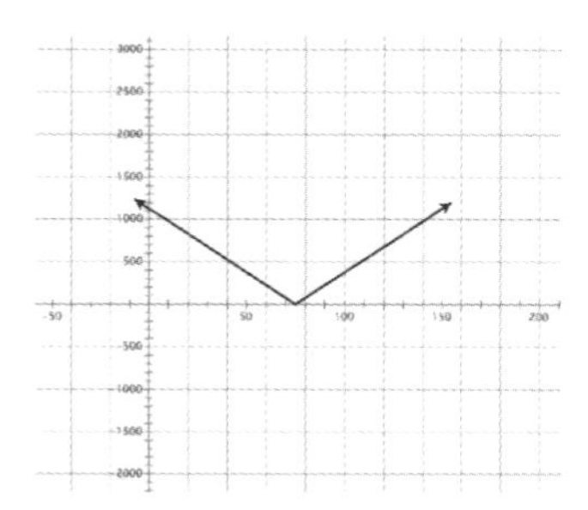

II.

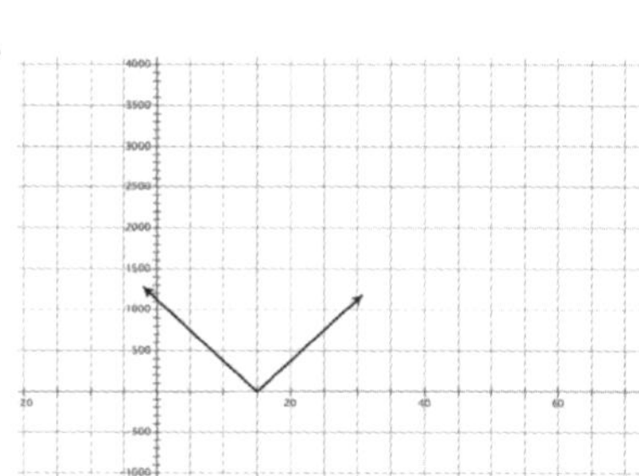

III.

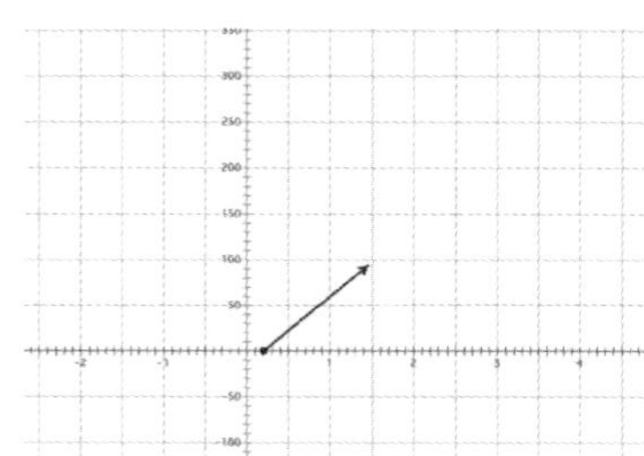

IV.

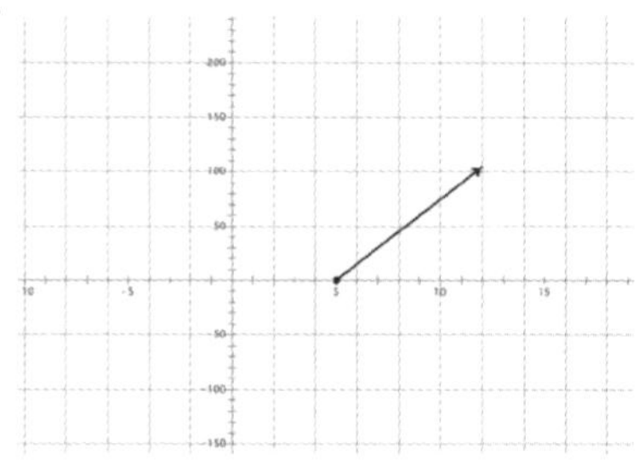

# Which Exponential Function Is It?

Name: ____________________ Date: ____________________

**1.** Could the function that represents the data in this table be an exponential function? Explain your reasoning.

| x | f(x) |
|---|---|
| 1 | 12 |
| 2 | 72 |
| 3 | 432 |

**2.** Exponential functions can have equations in the form $f(x) = Ab^x$. We know $f(1) = 12$, $f(2) = 72$, and $f(3) = 432$.

a. Replace $f(1)$ and $f(2)$ with expressions involving $A$ and $b$ by using the formula $f(x) = Ab^x$.

b. Solve for $A$ and $b$ so that both equations you found in part a are true. Now, you have a specific formula for $f(x)$, if you substitute the values of $A$ and $b$ you found.

c. Check that $f(3) = 452$ using the formula for $f(x)$ you found in part b.

**3.** Use the method you developed in item 2 to find $A$ and $b$ so that $g(x) = Ab^x$ for this function.

| x | g(x) |
|---|---|
| 2 | 3 |
| 4 | 12 |
| 7 | 96 |
| 11 | 1,536 |

# Coupon Stacks

Name: ______________________________ Date: ____________________

Nik is the popular owner of MathStyles, which specializes in mathematics-themed T-shirts. He really likes a particular grape soda brand, so his employees bring him coupons whenever they find them, which he keeps on the corner of his desk. On January 1, Nik found one coupon on his desk. One week after January 1 there were two coupons, 2 weeks after January 1 there were a total of four coupons, and 3 weeks after January 1 there were a total of eight coupons.

**1.** If this pattern (each week there are twice as many coupons as the week before) continues, how many coupons will be on his desk:

a. 10 weeks after January 1?

b. 26 weeks after January 1?

**2.** About how many miles high would the total stack be after 26 weeks? How might you determine the height? Is this possible?

a. How many tons would the coupon stack weigh after 26 weeks? How might your determine the tonnage?

b. How reasonable is this scenario? Justify your response.

# Tea with Mrs. Wiley

Name: ______________________________ Date: ______________________

Steve's favorite mathematics teacher regularly came to the café that Steve owned. Every Monday morning, Mrs. Wiley arrived at exactly 9:15 with her daughter Jackie. Jackie insisted on drinking freshly brewed tea but would only drink it if its temperature was *exactly* 85° Fahrenheit (and she made a loud scene if the tea was too hot or too cold!).

Steve had learned about Newton's Law of Cooling and decided to use it to his advantage by determining exactly when to boil the water so that Jackie's tea would be 85° at 9:15. Newton's Law of Cooling says that the temperature of the tea will be modeled by the function

$$f(t) = A \times 2^{-kt} + R,$$

where $t$ is the time in minutes because the hot water was poured into the cup, $R$ is the temperature of the surrounding air, and $A + R$ is the temperature of the hot water when it is poured into the cup. The number $k$ depends on the insulating properties of the tea cup that is being used.

1. Steve knew that the temperature of the café was always kept at 75°F and that the tea was brewed with water at the boiling point (212°F). Use this information to determine $R$ and $A$ in the formula above and rewrite the formula below by filling in the blanks.

$$f(t) = \_\_\_\_ \times 2^{-kt} + \_\_\_\_$$

2. Now, we need to help Steve determine the value of $k$. One Monday morning, Steve boiled the water for Jackie's tea at exactly 9:05, but

when he served it at 9:15, the tea's temperature was 100°F and Jackie loudly complained. That afternoon, Steve did some calculating and found that if the tea cooled down to 100°F in 10 minutes, then $k$ must be approximately -0.245. Show that if you substitute $k = -0.245$ into your formula for $f(t)$, then $f(10)$ is approximately 100.

**3.** Now, determine the time Steve should brew Jackie's tea by using guess-and-check and your formula for $f(t)$ to determine how long (to the nearest half minute) it will take the tea to cool to 85°F. Record your trials in a table.

# Piecing Together the Cost

Name: ______________________________ Date: ______________________

**1.** The U.S. Postal Service has a special "library rate" to mail books. A partial list of prices (in dollars) as a function of weight (in pounds) is given by the piecewise function $C(x)$ to the right. Fill in the table below with the appropriate costs.

$$C(x) = \begin{cases} 2.56 & \text{if } 0 < x \leq 1 \\ 3.02 & \text{if } 1 < x \leq 2 \\ 3.48 & \text{if } 2 < x \leq 3 \\ 3.94 & \text{if } 3 < x \leq 4 \\ 4.40 & \text{if } 4 < x \leq 5 \\ 4.86 & \text{if } 5 < x \leq 6 \\ 5.32 & \text{if } 6 < x \leq 7 \\ 5.76 & \text{if } 7 < x \leq 8 \\ 6.20 & \text{if } 8 < x \leq 9 \\ 6.64 & \text{if } 9 < x \leq 10 \end{cases}$$

| **Weight (in pounds)** | 0.75 | 1.283 | 5.001 | 6.999 | 8.7 |
|---|---|---|---|---|---|
| **Cost (in dollars)** | | | | | |

**2.** Fill in the blanks below to complete the description of the costs of mailing books via U.S. mail.

The cost of mailing a book via the U.S. Postal Service is $________ for packages weighing up to 1 pound. Every additional pound (or fraction of a pound) up to a total weight of _______ pounds costs _______ cents. Beyond a total of _______ pounds, every additional full (or fraction of) a pound costs ________ cents.

**3.** Alpha Wireless and Beta Cellular are two competing cell phone service providers specializing in non-smartphone users. The costs per minute (in dollars) of each company are given below as piecewise functions.

$$A(x) = \begin{cases} 40 & \text{if } 0 \leq x \leq 500 \\ 40 + .05\,(500 - x) & \text{if } x > 500 \end{cases}$$

Alpha Wireless

$$B(x) = \begin{cases} 25 & \text{if } 0 \leq x \leq 300 \\ 25 + .1\,(300 - x) & \text{if } x > 300 \end{cases}$$

Beta Cellular

a. Fill in the blanks to complete the written descriptions of the two plans:

Alpha Wireless service costs $_______ for up to ________ minutes of use plus an additional _________ cents for every minute over _________ minutes.

Beta Cellular service costs $_________ for up to ________ minutes of use plus an additional _________ cents for every minute over _________ minutes.

b. Sophia is very consistent with her cell phone use, always using between 425 and 475 minutes each month. Analyze the two cell phone plans and determine whether Sophia should choose Alpha Wireless or Beta Cellular. Explain your reasoning.

# Finite Differences Revisited 1

Name: ______________________________ Date: ____________________

1. This table shows some input-output pairs for the polynomial function $f(x)$. Guess a value for $f(6)$, being sure to explain your reasoning.

| x | f(x) |
|---|---|
| 2 | 7 |
| 3 | 11 |
| 4 | 15 |
| 5 | 19 |
| 6 | |

2. What kind of function does $f(x)$ seem to be? What degree is the polynomial that represents $f(x)$ (that is, what is the highest power of $x$ that appears in the formula)? Find a formula for $f(x)$. Explain your reasoning.

3. Fill in the outputs and differences in this table for the linear function $(x) = Mx + B$.

| x | f(x) | Difference: f(x+1)– f(x) |
|---|---|---|
| | | |
| 1 | | |
| 2 | | |
| 3 | | |
| 4 | | |
| 5 | | |

4. Write a formula for a function that matches the information in each table. Explain your work.

| x | f(x) |
|---|---|
| 1 | 3 |
| 2 | 9 |
| 3 | 15 |
| 4 | 21 |

| x | f(x) |
|---|---|
| 1 | 9 |
| 2 | 5 |
| 3 | 1 |
| 4 | -3 |

# Finite Differences Revisited 2

Name: ______________________________ Date: ______________________

In this activity, you will find ways to recognize quadratic polynomials represented in input-output tables.

1. A table of values, including outputs, first differences, and second differences, has been started for the quadratic function $g(x) = Ax^2 + Bx + C$. Complete the next two rows in each column of the table.

| x | g(x) | DIFF(x) = g(x + 1) – g(x) | DIFF(x + 1) – DIFF(x) |
|---|---|---|---|
| | | | |
| 1 | $A + B + C$ | $(4A + 2B + C) - (A + B + C) = 3A + B$ | $(5A + B) - (3A + B) = 2A$ |
| 2 | $4A + 2B + C$ | $(9A + 3B + C) - (4A + 2B + C) = 5A + B$ | $(7A + B) - (5A + B) = 2A$ |
| 3 | $9A + 6B + C$ | $(16A + 4B + C) - (9A + 3B + C) = 7A + B$ | |
| 4 | $16A + 4B + C$ | | |
| 5 | | | |
| 6 | | | |

2. Fill in the row corresponding to $x = 0$ (the blank row near the top of the table).

3. We noticed before that if $f(x)$ is a linear function, then DIFF($x$) is constant. Is that the case for $g(x) = Ax^2 + Bx + C$? Is anything else constant?

4. Kyrie was looking at the input-output table that follows and said she had a way of finding a formula that matched the values in the table. "Once I saw that the second differences were constant, I realized I could just use the $x = 1$ row to find the coefficients of a quadratic polynomial that matched the table. I can figure out $A = 2$ by noticing that DIFF($x + 1$) – DIFF($x$) is always $2A$. Then, because $g(1) - g(0) = 3A + B$, I can figure out that $B = -4$. Now that I know $A$ and $B$, I can use $g(1) = A + B + C$ to find that $C = 3$. Just to make sure, I checked that all the given values matched the table if $g(x) = 2x^2 - 4x + 3$."

| x | g(x) | DIFF(x) = g(x + 1) – g(x) | DIFF(x + 1) – DIFF(x) |
|---|---|---|---|
| | | | |
| 1 | 1 | | |
| 2 | 3 | | |
| 3 | 9 | | |
| 4 | 19 | | |
| 5 | 33 | | |
| 6 | 51 | | |

Use Kyrie's method (or one of your own) to find a possible formula for the function with the following input-output table. Be sure to explain your reasoning.

| x | g(x) | DIFF(x) = g(x + 1) – g(x) | DIFF(x + 1) – DIFF(x) |
|---|---|---|---|
| | | | |
| 1 | 3 | | |
| 2 | 4 | | |
| 3 | 7 | | |
| 4 | 12 | | |
| 5 | 19 | | |
| 6 | 28 | | |

**5.** What about this one?

a. In the table that follows, compute the outputs for this function:

$$f(x) = x^6 - 21x^5 + 175x^4 - 735x^3 + 1{,}625x^2 - 1764x + 720$$

Use your calculator if necessary. It will help to know that another representation of $f(x)$ is:

$$f(x) = x^2 + (x-1)(x-2)(x-3)(x-4)(x-5)(x-6)$$

| x | f(x) |
|---|---|
| 1 | |
| 2 | |
| 3 | |
| 4 | |
| 5 | |
| 6 | |
| | |

a. What less complicated function would you have *predicted* the table was for?

c. Why does it make sense that the less complicated function and $f(x)$ agree on the inputs $x = 1, 2, 3, 4, 5,$ and $6$?

d. Show that the output of the less complicated function when $x = 7$ does *not* equal $f(7)$.

# ANSWER KEY

## Expressions

### Classroom Expressions A1-2

**1.** -19
**2.** 18
**3.** 7
**4.** 16
**5.** -1
**6.** 24
**7.** 32
**8.** 1
**9.** -62
**10.** 11
**11.** 1
**12.** 0.6

### In What Order? A1-3

**1.** $6 - 8 + 3 \times 4$
$6 - 8 + 12$; multiplication
$-2 + 12$; subtraction
10; addition

**2.** $7 + 6(8 - 4 + 3) + 6 \cdot 3$
$7 + 6(4 + 3) + 6 \cdot 3$; parenthesis, subtraction
$7 + 6(7) + 6 \cdot 3$; parenthesis, addition
$7 + 42 + 18$; multiplication
$49 + 18$; addition
67; addition

**3.** $3^2 + 4(2 - 5 \cdot 3) \div 14 - 12$
$3^2 + 4(2 - 15) \div 14 - 12$; parenthesis; multiplication
$3^2 + 4(-13) \div 14 - 12$; parenthesis, subtraction
$9 + 4(-13) \div 14 - 12$; exponents
$9 - 52 \div 14 - 12$; multiplication
$9 - 3\frac{5}{7} - 12$; division
$5\frac{2}{7} - 12$; addition
$-6\frac{5}{7}$; subtraction

**4.** Madison is correct; $-2^2$ means raise the 2 to the second power and then multiply by -1.
$-4 - 3(12 - 5) \div 3 \times -2^2$
$-4 - 3(7) \div 3 \times -4$
$-4 - 21 \div 3 \times -4$
$-4 - 7 \times -4$
$-4 + 28$
24

### Equivalent Expressions A1-6

The cards are shown in the correct order on the original. Start at top left and go to the right. At the end of each row, go to the leftmost entry of the next row until you get to the simplified expression. The remaining cards in the set are incorrect answer choices.

### What's My Expression? A1-9

**1.** Multiply the input by three; then add two; $3n + 2$.

**2.** Multiply the input by negative 6; then add two; $-6n + 2$.

**3.** Square the input then multiply by -4; $-4n^2$.

**4.** It is not possible to write a linear or quadratic expression with these data. Discuss with your students why it is not possible to write either the linear or quadratic expression. You might also ask them to graph the data to see that the line is not straight or a parabola.

### Match the Equivalents A1-10

| W | E |
|---|---|
| **Add six to *n*; then multiply by two.** | $2(n + 6)$ |
| **Square the quantity of *n* plus four.** | $(n + 4)^2$ |
| **Multiply *n* by two; then add six.** | $2n + 6$ |
| **Square *n*; then add six.** | $n^2 + 6$ |
| **Write the factors of 4*n* – 6.** | $2(2n - 3)$ |
| **Divide the quantity *n* plus six by two.** | $\frac{n + 6}{2}$ |
| **Add six to *n*, then square the sum.** | $(n + 6)^2$ |
| **Divide *n* by two, then add six.** | $\frac{n}{2} + 6$ |
| **Raise negative four to the second power.** | $(-4)^2$ |
| **Multiply the product of four times *n* by negative one.** | $-(4n)$ |

## Growing Figures 1 A1-12

$$4(n - 1) + 1$$

## What's My Rule? A1-13

**1.** Answers will vary. Sample response: Each figure has a square on the bottom left and top right (by itself). Between these squares, there is a rectangle with $n$ columns of $n + 2$ squares.

**2.**

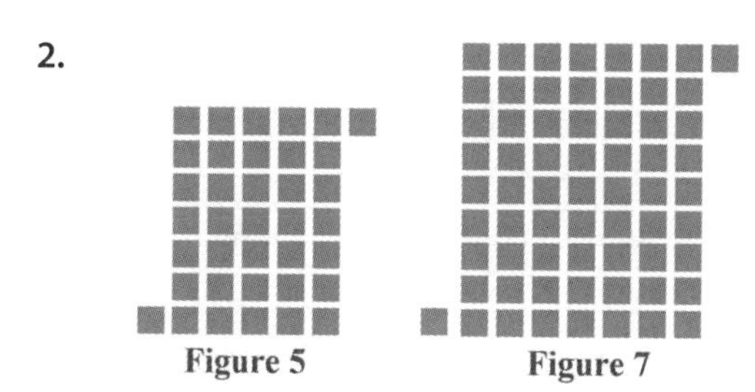

Figure 5 Figure 7

**3.** $n(n + 2) + 2$ *or* $n^2 + 2n + 2$

## Growing Figures 2 A1-14

**1.**

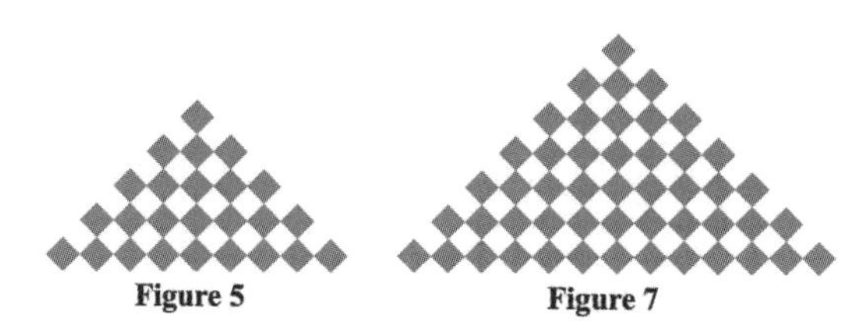

Figure 5 Figure 7

Figure $n$ has $n$ rows of rhombi. The numbers of rhombi in each row are consecutive odd numbers from 1 to $2n - 1$. The number of rhombi in Figure $n$ is $n^2$.

**2.**

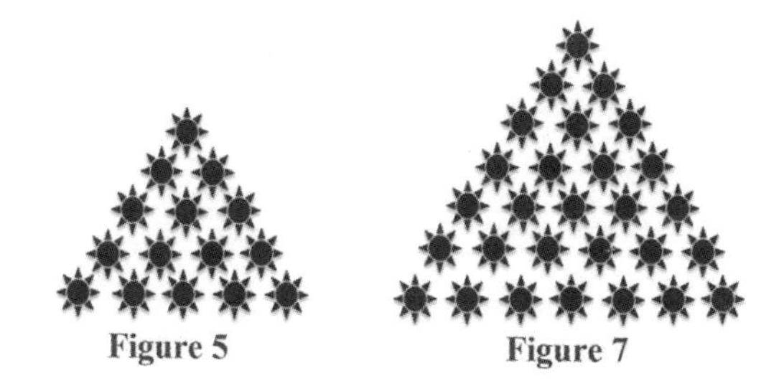

Figure 5 Figure 7

Figure $n$ has $n$ rows of stars. The numbers of stars in each row are the numbers from 1 to $n$ consecutively. The number of stars in Figure $n$ is $\frac{n(n+1)}{2}$, the sum of the counting numbers from 1 to $n$.

**3.**

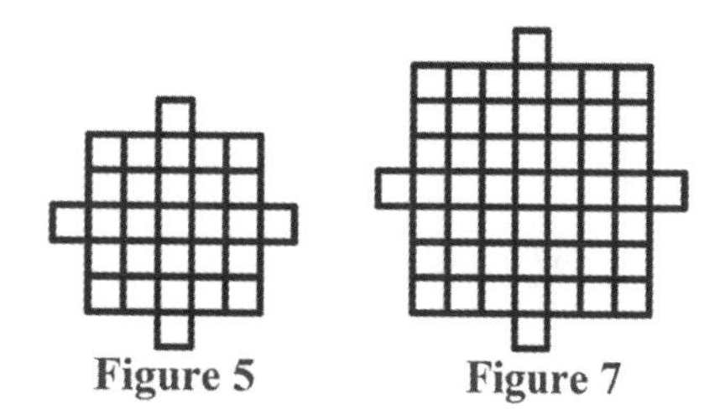

Figure 5 Figure 7

Figure $n$ has an $n$-by-$n$ square surrounded by 4 unit squares for a total of $n^2 + 4$ unit squares.

**4.**

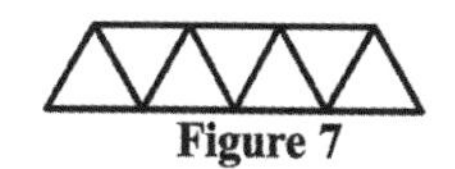

Figure 5 Figure 7

Figure $n$ is made up of $n$ equilateral triangles attached by edges. Every other one is upside down. The perimeter of figure $n$ is $n + 2$, assuming each edge has length 1.

## My Counterpart A1-15

The counterparts share a row in the original.

## How Much Will I Pay? A1-17

**1.** **a.** $900(1 + 0.06)^2$ or $900(1.06)^2$
**b.** $900(1.06)^2 = 900(1.1236) = \$1,011.24$

**2.** $850(1.08) = \$918.00$

**3.**

÷ 50

17 — 68 — \$50 — 918 — cost

×4 ×4

2 — 8 — 100 — 108 — Percent

÷ 50

**4.** After the 10% pay cut, the average pay was $55,000(0.90) = \$49,500$. After the 10% raise, it went up to $49,500(1.10) = \$54,450$ (*less* than they started with!).

**5.** Price before tax is $7(0.75) = \$5.25$. After tax is applied, the cost will be $5.25(1.06) = 5.565$, which rounds to \$5.57.

## Taxes, Interest, and More A1-18

Models may vary.

**1.** $499(0.07) = \$34.93$

| Value | Percent |
|---|---|
| 499 | 100 |
| 4.99 | 1 |
| | |
| 34.93 | 7 |

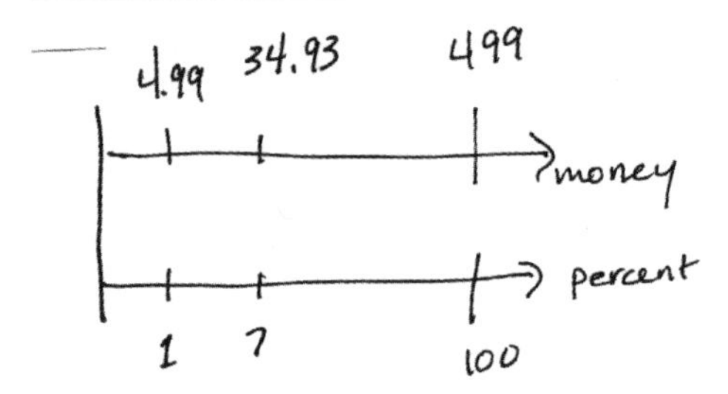


**2.** $119(1.055) = \$125.55$

| Value | Percent |
|---|---|
| 119 | 100 |
| 11.9 | 10 |
| 1.19 | 1 |
| 0.595 | 0.5 |
| 5.95 | 5 |
| 125.55 | 105.5 |

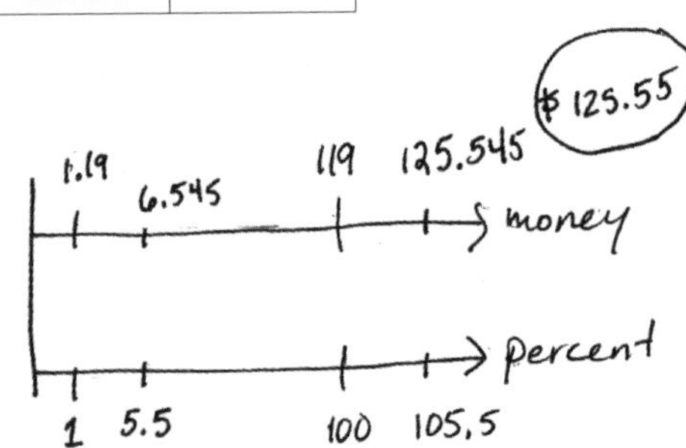


**3.** $88{,}400(0.03) = \$2{,}652$

| Value | Percent |
|---|---|
| 88,400 | 100 |
| 8,840 | 10 |
| 884 | 1 |
| 2,652 | 3 |

**4.** $6{,}500(1.015)^3 = \$6{,}796.91$

| Value | Interest | Value | Year |
|---|---|---|---|
| 6,500 | | | 0 |
| 6,500 | 97.50 | 6,597.50 | 1 |
| 6,597.50 | 98.96 | 6,696.46 | 2 |
| 6,696.46 | 100.45 | 6,796.91 | 3 |
| | | | |
| | | | |
| | | | |

## Venn Diagram Representations A1-20

Students may not write *1* in sections of the diagram with no variable. Accept responses that are otherwise correct, but help them see why including a 1 is correct and helpful.

**1.** $x^2y^3$ | $y$ | $x^2y^2$ | $x$ | $x^3y^2$ $a^4b^2$ | $a^2b$ | $a^2b$ | $1$ | $a^2b$

GCF $(x^2y^3, x^3y^2) \rightarrow x^2y^2$, GCF $(a^4b^2, a^2b) \rightarrow a^2b$

LCM $(x^2y^3, x^3y^2) \rightarrow x^3y^3$, LCM $(a^4b^2, a^2b) \rightarrow a^4b^2$

**2.** $c^5d^3$ | $c^3d$ | $c^2d^2$ | $1$ | $c^2d^2$ $x^4y^8$ | $y^6$ | $x^4y^2$ | $x^2$ | $x^6y^2$

GCF $(c^5d^3, c^2d^2) \rightarrow c^2d^2$, GCF $(x^4y^8, x^6y^2) \rightarrow x^4y^2$

LCM $(c^5d^3, c^2d^2) \rightarrow c^5d^3$, LCM $(x^4y^8, x^6y^2) \rightarrow x^6y^8$

**3.** $a^3b^4$ | $1$ | $a^3b^4$ | $ab$ | $a^4b^5$ $c^2d$ | $1$ | $c^2d$ | $cd^3$ | $c^3d^4$

GCF $(a^3b^4, a^4b^5) \rightarrow a^3b^4$, GCF $(c^2d, c^3d^4) \rightarrow c^2d$

LCM $(a^3b^4, a^4b^5) \rightarrow c^5d^3$, LCM $(c^2d, c^3d^4) \rightarrow c^3d^4$

## Factor Lattices 1 A1-21

**1.** Factor lattice for $x^5y^3$

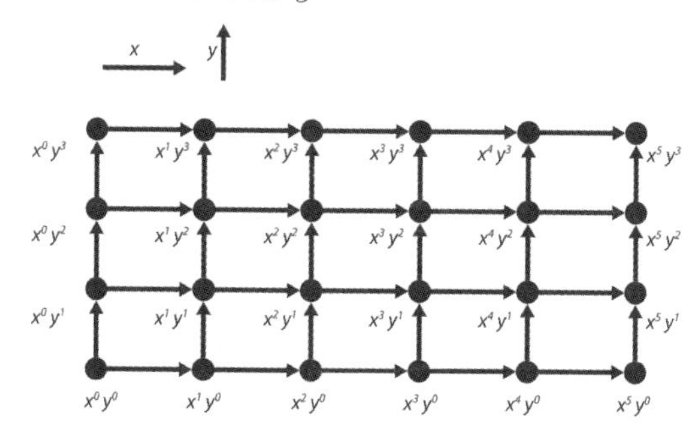

**a.** LCM $(x^2, y) \rightarrow x^2y$ **b.** LCM $(x^3, y^2) \rightarrow x^3y^2$

**2.** **a.** LCM $(xy, y^3) \rightarrow xy^3$ **b.** LCM $(x, x^2y^2)$ $\rightarrow xy^2$

**3.** **a.** GCF $(xy^3, x^2y^3) \rightarrow xy^3$ **b.** GCF $(x^2y^3, y^2)$ $\rightarrow y^2$

## Modeling Algebraic Expressions A1-22

**1.** The result is $b^2 - 2b$. The $b$-by-$b$ – 2 rectangle can be thought of as a $b$-by-$b$ square with two 1-by-$b$ rectangles or one 2-by-$b$ rectangle removed. Its area is $b^2 - 2b$ sq units.

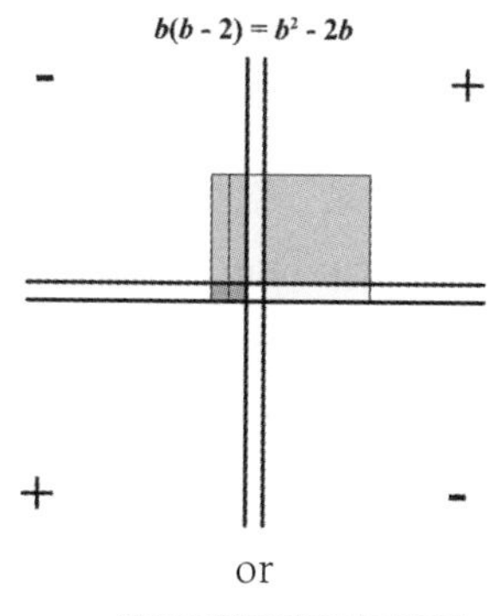



or

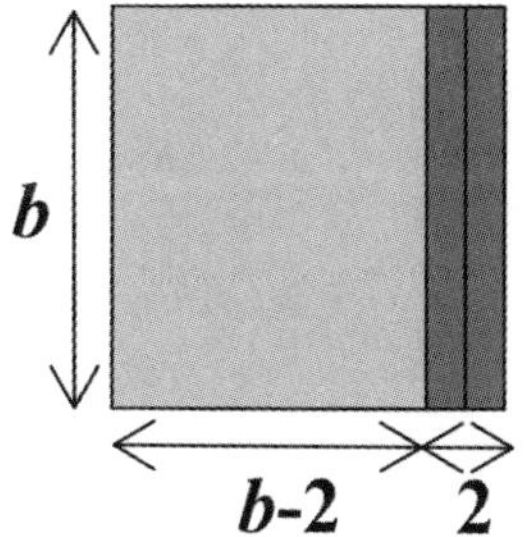



**2.** The result is $-(b^2 - 2b)$ or $-b^2 + 2b$ or $2b - b^2$.

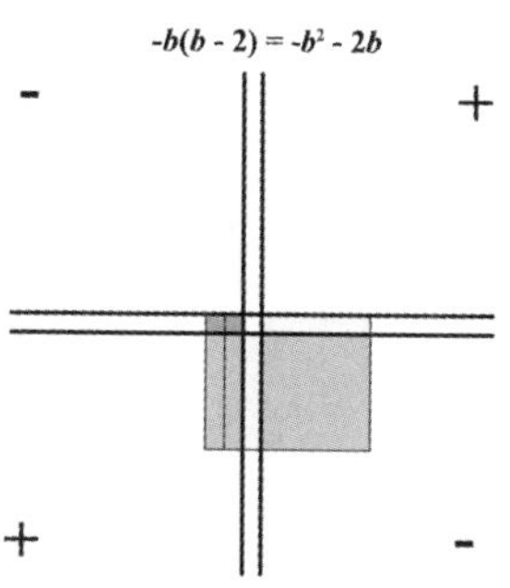



**3.** The result is $b^2 + 6b + 9$ sq units. The b + 3-by-$b$ + 3 square can be created by appending six 1-by-$b$ rectangles (three on top and three on the side) and a 3-by-3 square to the $b$-by-$b$ square. The resulting area is then $b^2 + 6b + 9$ sq units.

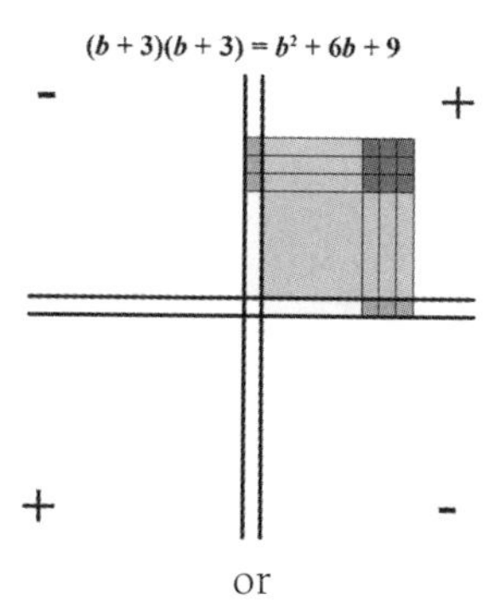



or

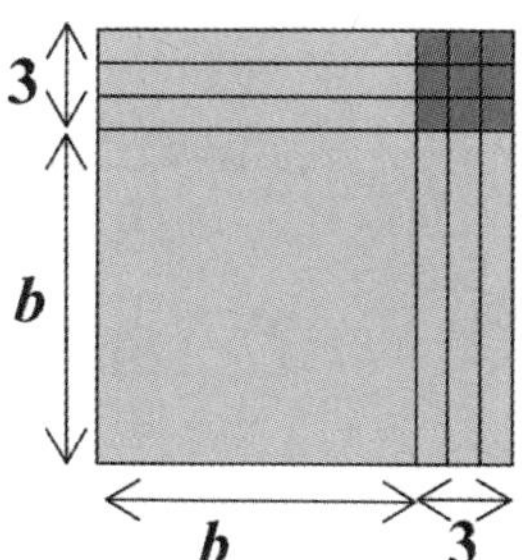



**4.** The result is $(b - 6)(b + 1)$. To get an area of $b^2$, we need a $b$-by-$b$ square. The $-5b$ will result from five 1-by-$b$ rectangles below or to the left of the square. To get a rectangle of area 6, we need a 2-by-3 or 1-by-6 rectangle. The 6-by-1 rectangle (containing six unit squares) works because a $b$ – 6-by-$b$ + 1 rectangle will have area $b^2 - 5b - 6$ sq units.

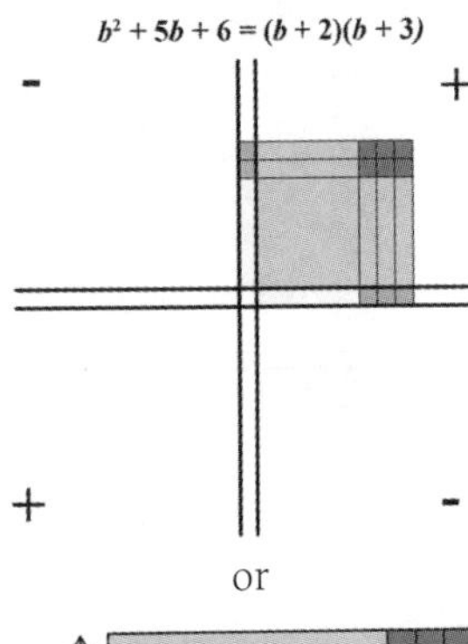


or

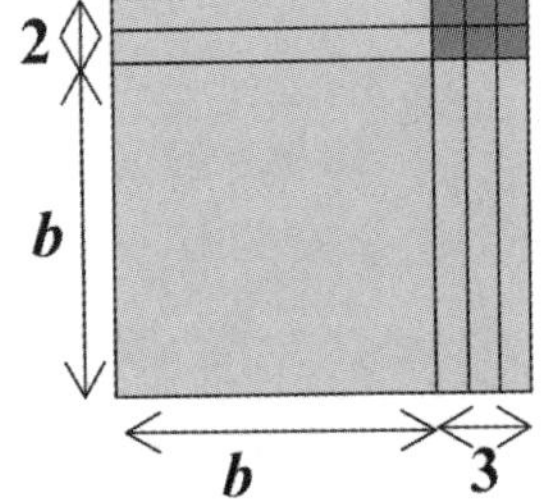


5. The result is $(b - 6)(b + 1)$. The $b - 6$-by-$b + 1$ rectangle is obtained by starting with a $b$-by-$b - 6$ (with area $b^2 + b$), then removing one 1-by-$b$ rectangles and six unit squares, resulting in an area of $b^2 + b - 6b - 6 = b2 - 5b - 6$.

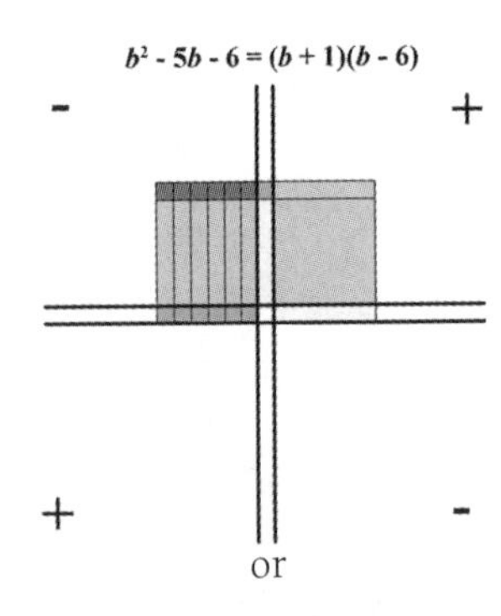


or

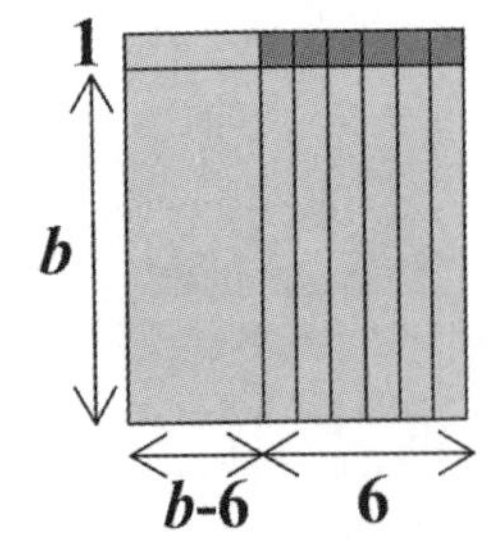


6. The result is $(b - 4)(b - 2)$. Starting with a $b$-by-$b$ square (with area $b^2$), subtract an area of $6b$, which can be accomplished by subtracting two $b$-by-1 rectangles from the top and four $b$-by-1 rectangles from the right of the figure. Adding an area of 8 corresponds to the eight unit squares in the overlap of the earlier rectangles (which was subtracted twice), so the resulting area is the same as the area of the $b - 4$-by-$b - 2$ rectangle.

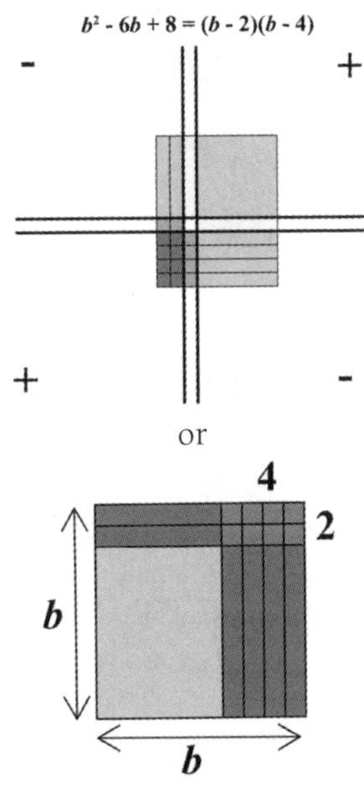


7. The result is $b^2 - 4b + 4$. If you start with the $b$-by-$b$ square and remove four 1-by-$b$ rectangles (two on top and two on the side), then the 2-by-2 square at the top left of the figure has been removed twice (and we only wanted to remove it once). Adding it back in gives us the $b - 2$-by-$b - 2$ square, which has the same area as $b^2 - 4b + 4$.

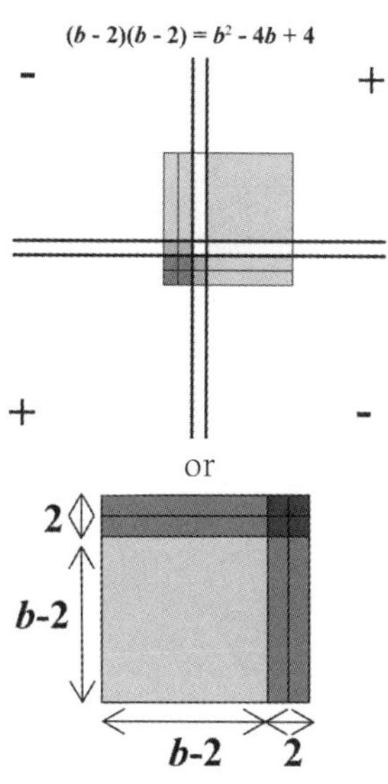


**8.** The result is $(b - 8)(b + 3)$. To subtract twenty-four, we need twenty-four unit squares, so we need to find two numbers whose product is twenty-four and whose difference is five. Starting with the $b$-by-$b$ square, we subtract $8b$ (the eight 1-by-$b$ rectangles on the lower right, then add $3b$ (the three $b$-by-1 rectangles on the top), then subtract the twenty-four unit squares (the 8-by-4 rectangle on upper right), leaving the $b - 8$-by-$b + 3$ rectangle, which has area $b^2 - 5b - 24$ sq units.

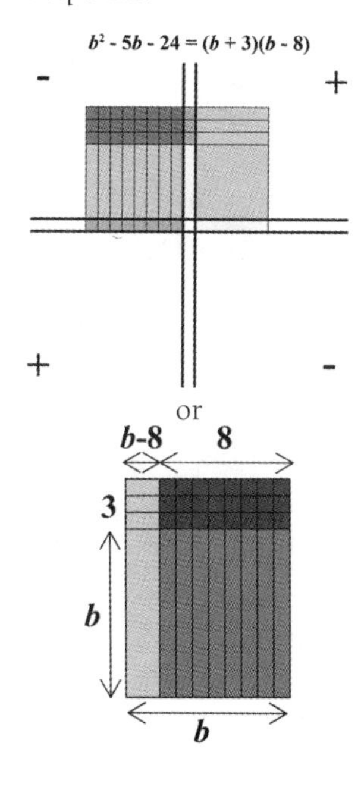


**9.** The result is $b^2 - 9$ sq units. The $b + 3$-by-$b - 3$ rectangle is the darkly shaded region in the figure. The darkly shaded $b$-by-3 rectangles on the right side of the figure has the same area as the light gray $b$-by-3 rectangles at the top left. The dark part of the "right" rectangles is nine unit squares smaller than the three gray rectangles. That means that the area of the $b$-by-$b$ square ($b^2$) is nine units more than the area of the $b + 3$-by-$b - 3$ rectangle, so $(b + 3)(b - 3) = b^2 - 9$ sq units.

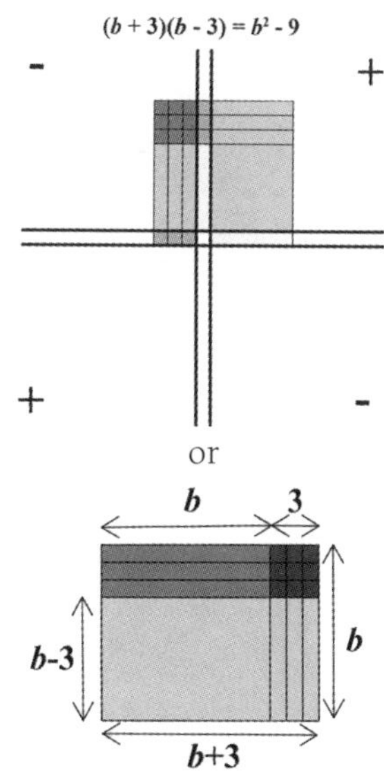


**10.** The result is $(b - 4)(b + 4)$. The area of the figure created by removing a 4-by-4 square from the $b$-by-$b$ square is $b^2 - 16$. The rectangle at the top left of the figure has the same dimension (4 by $b - 4$) as the one on the lower right. If the top left rectangle is rotated and moved next to the lower right rectangle, the new figure will be a $b - 4$-by-$b + 4$ rectangle, which has an area equal to $(b - 4)(b + 4)$.

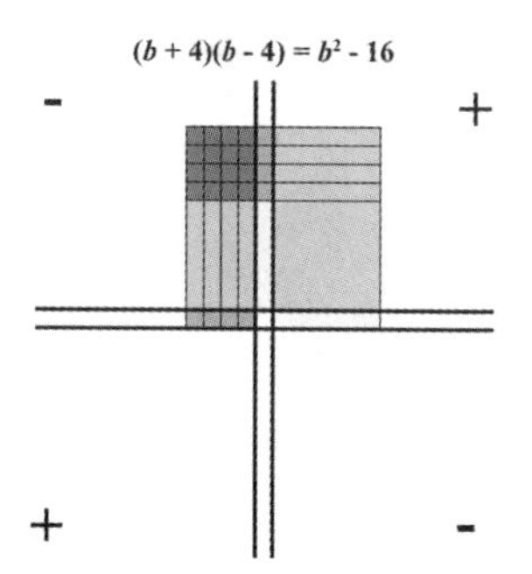

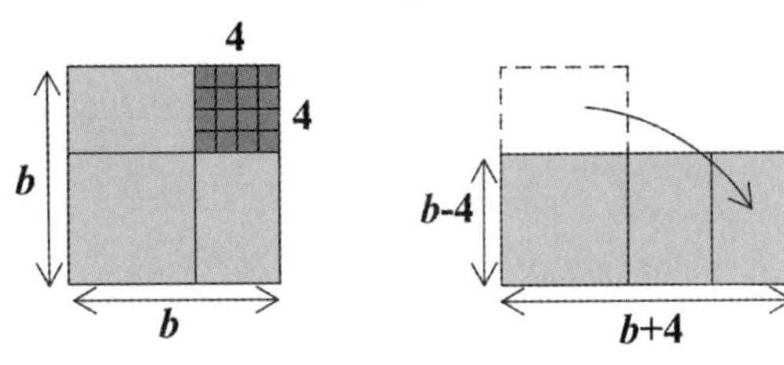


## Expanding Products 1 A1-23

**1.** $a^2 + 5a + 6$

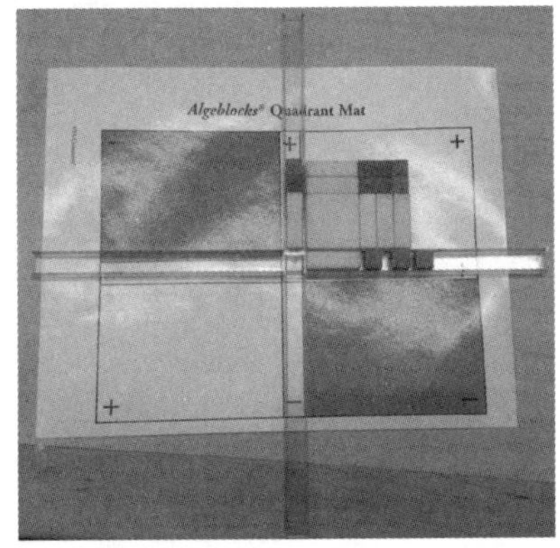


**2.** $a^3 + 3a^2 + 3a + 1$. Notice that one factor is linear $a + 1$ and the second factor is two dimensional $(a^2 + 2a + 1)$. The degree of the polynomial is 3 so the figure is three dimensional.

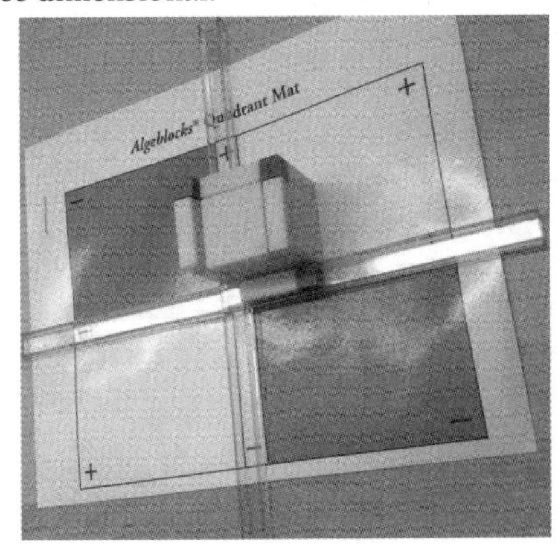

**3.** $a^3 + 8a^2 + 21a + 18$. The degree of the polynomial is 3 so the figure is three dimensional.

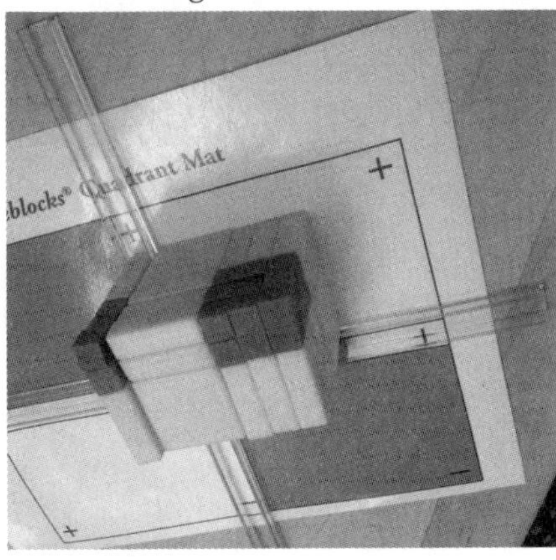

**4.** $a^4 + 10a^3 + 37a^2 + 60a + 36$. The degree of this polynomial is 4, so the figure is four dimensional, which is not possible to model using Algeblocks. $(a + 3)^2$ is equivalent to $a + 6a + 9$ and $(a + 2)^2$ is equivalent to $a^2 + 4a + 4$. The product of these trinomials is $a^4 + 10a^3 + 37a^2 + 60a + 36$.

## Expanding Products 2 A1-24

**1.** The length and width of the box are $(a + 3)$ inches and $(a + 1)$ inches.

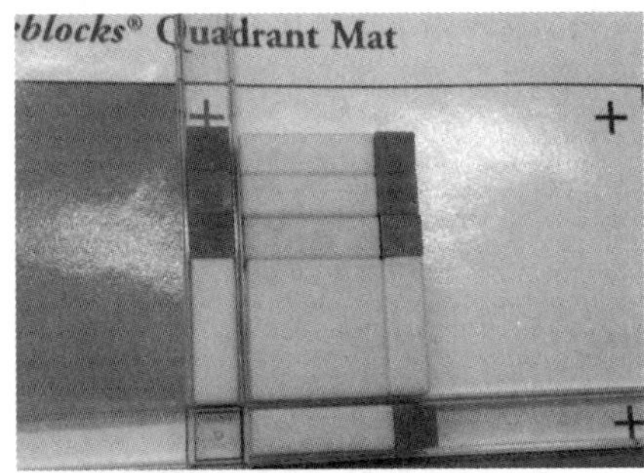


**2.** The dimensions of the plot of land are $(4x + 10)$ feet by $(4x + 2)$ feet.

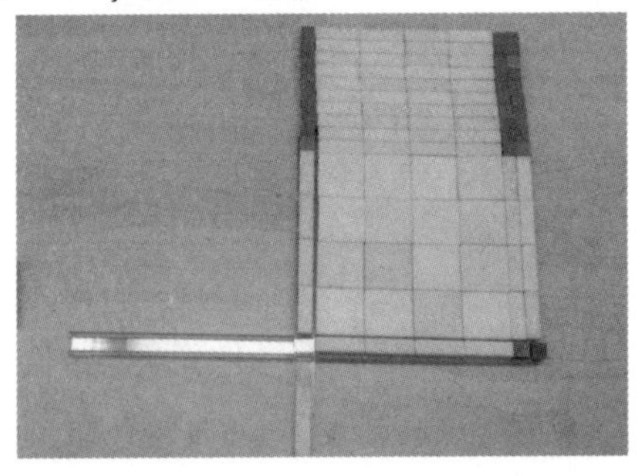

3. Given a height of $x + 1$ feet, the other two dimensions are $x + 2$ feet and $x + 3$ feet. A container with volume $x^3 + 6x^2 + 11x + 6$ cubic feet and height $x + 1$ feet has a length and width whose product is $x^2 + 5x + 6$ square feet. Looking at the solution of item 1, we can see that the length and width must be $x + 3$ feet and $x + 2$ feet.

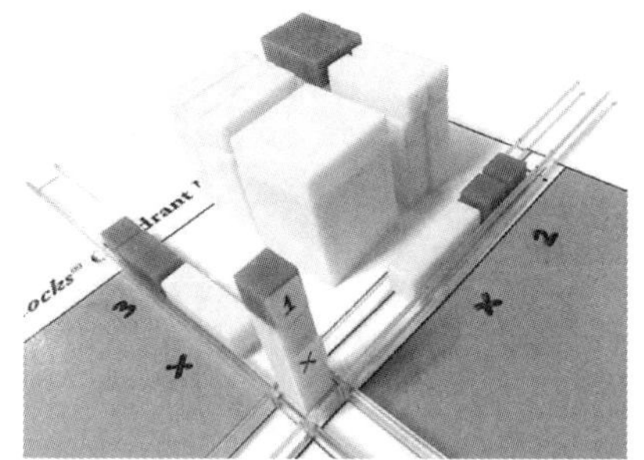

## Don't Get FOILed Again! A1-25

1.

a. Increase each dimension by 6 feet; $(x + 6)$ feet by $(x + 6)$ feet.

b. Area = $x^2 + 12x + 36$ square feet

2.

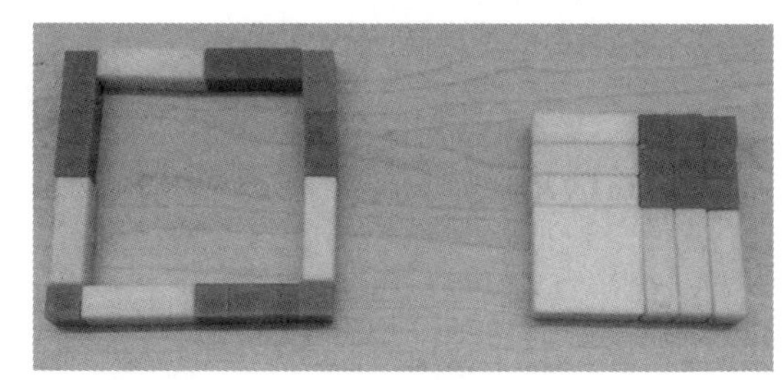

a. Deion's original square is now $3x$ feet wider and $3x$ feet longer, which means that 9 square feet must be added in the corner to complete the square. The area of the new walkway is $x^2 + 6x + 9$ square feet.

b. The walkway is really a "hole" filled with 6 inches (or ½ foot) of pea stone. This means that the volume of the pea stone is $\frac{1}{2}(x^2 + 6x + 9)$ cubic feet.

3. $24x^3 - 58x^2 + 23x + 15$

To see that this is the answer, carefully expand the product (it's not really feasible to model the final answer with Algeblocks):

$(4x - 5)(3x + 1)(2x - 3)$

$= (4x - 5)((3x)(2x) + 1(2x)+3x(-3) + 1(-3))$

$= (4x - 5)(6x^2 - 7x - 3)$

$= ((4x)(6x^2 - 5(6x^2 )+(4x)(-7x) - 5(-7x) + (4x)(-3) - 5(-3)$

$= 24x^3 - 58x^2 + 23x + 15$

## Operating with Exponents A1-26

1. a. If you expand each term, you get $2 \times 2 \times 2 \times 2 \times 2 \times 2 \times 2 \times 2$, or $2^8$.

b. If you expand each term, you get $5 \times 5 \times 5 \times 5 \times 5 \times 5 \times 5 \times 5 \times 5 \times 5 \times 5$, or $5^{11}$.

2. $a^{m+n}$: When the bases are the same, you keep the base and add the exponents.

3. a. If you expand each term, you get $2 \times 2 \times 2 \times 3 \times 3 \times 3$. Use the commutative property to write the equivalent expression $2 \times 3 \times 2 \times 3 \times 2 \times 3$, which is equivalent to $6^3$.

b. $5 \times 5 \times 5 \times 5 \times 5 \times 5 \times 5 \times 6 \times 6 \times 6 \times 6 \times 6 \times 6 \times 6 = 5 \times 6 \times 5 \times 6 \times 5 \times 6 \times 5 \times 6 \times 5 \times 6 \times 5 \times 6 \times 5 \times 6 = 30^7$

4. $(ab)^m$: When the bases are different and the exponents are the same, multiply the bases and raise them to the given exponent.

5. a. If $4^2$ is decomposed to its prime factors, the result is $(2^2)^2$, which is $2^{2\times2} = 2^4$.

b. If $9^2$ is decomposed to its prime factors, the result is $(3^2)^2$, which is $3^{2\times2} = 3^4$.

c. If $125^2$ is decomposed to its prime factors, the result is $(5^3)^2$, which is $5^{3\times2} = 5^6$.

6. $(a^m)^n$; To simplify raising a number to a power to a power, keep the base and multiply the exponents.

7. **a.** Because ½ = $2^{-1}$, this equals $(2^{-1})(2^n) = 2^{-1+n} = 2^{n-1}$.

**b.** Because $25 = 5^2$, this equals $(5^2)(5^{-2n}) = 5^{2-2n}$.

**c.** Because $4 = 2^2$, this equals $(2^2)^{n-1} = 2^{2(n-1)} = 2^{2n-2}$.

## Powerful Integers A1-27

1. **a.** Three raised to the power three; $3^3$; 27

**b.** The quantity negative five raised to the power four; $(-5)^4$; 625

**c.** $b$ raised to the power seven; $b^7$

2. **a.** Twelve raised to the power three; $12^3$; 1,728

**b.** Ten raised to the power negative five, or one divided by ten raised to the power five; $10^{-5}$; $(\frac{1}{100{,}000})$; $\frac{1}{10^5}$

**c.** $mn$ raised to the power four; $(mn)^4$

**d.** Written rule: multiply the bases and raise to the common power

3. **a.** Five raised to the power five; $5^5$; 3,125

**b.** Three raised to the negative two power, or one divided by three raised to the power two; $3^{-2}$; $(\frac{1}{3})^2$; $\frac{1}{9}$

**c.** The quantity negative $n$ raised to the power twelve; $(-n)^{12}$; $n^{12}$

**d.** Written rule: keep the base and add the exponents

4. **a.** Three raised to the power three; $3^{5-2} = 3^3$; 27

**b.** Five raised to the power negative four, or one divided by five raised to the power 4; $5^{3-7} = 5^{-4}$ or $\frac{1}{5^4}$; $\frac{1}{625}$

**c.** Six raised to the power negative eleven or one divided by six raised to the power eleven; $6^{-4-7} = 6^{-11}$ or $\frac{1}{6^{11}}$; $\frac{1}{362{,}797{,}056}$

**d.** $n$ squared or $n$ raised to the power two; $n^{5-3} = n^2$

**e.** Written rule: keep the base and subtract the exponents

5. **a.** $\frac{1}{3^4}$

**b.** $4^3$

**c.** $\frac{1}{m^5}$

## Where Do I Go? A1-29

1. Tamara might have had to walk five blocks east or five blocks west to get to the yogurt shop. If she walked five blocks east, she will be 3 + 5 = 8 blocks east of her house. If she walked five blocks west, she would have ended up 5 – 3 = 2 blocks west of her house.

2. If Parker and Lucas walked in the same direction, then Lucas walked 13 – 7 = 6 blocks past where Parker stopped, so they were six blocks apart. If Parker and Lucas walked in opposite directions, they were 7 + 13 = 20 blocks apart.

3. Using the west goal as the reference point, Sam started 15 yards east of the west goal and Kaylee started 95 yards east of the west goal. They ended up 50 yards east of the west goal, so Kaylee traveled 95 – 50 = 45 yards and Sam traveled 50 – 15 = 35 yards.

## Absolutely! A1-30

1. $|7 - (-10)|$ = 17 blocks, or $|7| + |-10|$ = 17 blocks

2. Edwin walks $|-20|$ = 20 blocks. Jasmin walks $|5|$ = 5 blocks. Edwin walks farther.

3. Jen-Min dove $|-60|$ = 60 feet below sea level.

## Broken Calculator A1-31

1. Because 7 is between 4 and 9, we know that $\sqrt{7}$ is between 2 and 3. Estimating $\sqrt{7}$ as 2.5 is close but a little too small because $2.5^2 = 6.25$ (multiplying 2.5 × 2.5 "by hand"). Because $2.6^2 = 676$ and $2.7^2 = 729$, we know that $\sqrt{7}$ is between 2.6 and 2.7. In fact, 2.65 is very close to $\sqrt{7}$ because $2.65^2 = 7.0225$.

2. Because 29 is between 25 and 36, we know that $\sqrt{29}$ is between 5 and 6. Moreover, we expect that $\sqrt{29}$ is closer to 5 than to 6. Because $5.5^2 = 30.25$, we know that $\sqrt{29}$ is a less than 5.5. Because $5.4^2 = 29.16$ and $5.3^2 = 28.09$, we see that $\sqrt{29}$ is between 5.3 and 5.4, closer to 5.4.

Because $5.38^2 = 28.9444$ and $5.39^2 = 29.0521$, we see that $\sqrt{29}$ is between 5.38 and 5.39, so 5.4 is a very good approximation.

3. We saw in the approximation of $\sqrt{29}$ that $5.5^2 = 30.25$, so $\sqrt{30.25}$ equals 5.5. We don't need to approximate it because we can compute its exact value.

4. Because $7^2 = 49$, we know that $\sqrt{51}$ will be very close to 7. We have $7.1^2 = 50.41$ and $7.2^2 = 51.84$, so $\sqrt{51}$ is between 7.1 and 7.2. Because $7.15^2 = 51.1225$ and $7.14^2 = 50.9796$, we see that 7.14 is a pretty good estimate for $\sqrt{51}$.

## Babylonian Square Roots A1-32

1. The white square measures 5 by 5, so area $5 \times 5 = 25$ square units. The two light gray 5-by-$\frac{1}{5}$ rectangles each have area $5 \times \frac{1}{5} = 1$ square unit, and the dark $\frac{1}{5}$-by-$\frac{1}{5}$ square has area $\frac{1}{5} \times \frac{1}{5} = \frac{1}{25}$ unit. The total area of the $5\frac{1}{5}$-by-$5\frac{1}{5}$ square is therefore $5\frac{1}{5} \times 5\frac{1}{5} = 25 + 1 + 1 + \frac{1}{25} = 27\frac{1}{25}$ square units.

2. We want to find a square whose area is approximately 68 square units. Its side length will then be approximately the square root of 68. Because 68 is a little more than $64 = 8^2$, we know that the square root of 68 is a little more than 8. We start with an 8-by-8 square, which has area 64 square units, which is 4 square units too small. That is, the 8-by-8 square is the right shape (a square), but the wrong area. Because the square is 4 square units too small, we add 4 square units on the right side of the square. To have area 4, the rectangle should have dimensions 8 by $\frac{1}{2}$. Now, we have an 8-by-$8\frac{1}{2}$ rectangle, which has the correct area, but the wrong shape (it's not a square). If we split the 8-by-$\frac{1}{2}$ rectangle into two 8-by-$\frac{1}{4}$ rectangles, keeping one on the right of the square and placing the other at the bottom of the square, we have an $8\frac{1}{4}$-by-$8\frac{1}{4}$ square, except for the tiny $\frac{1}{4}$-by-$\frac{1}{4}$ square at the bottom right of the figure. This shows that $8\frac{1}{4} \times 8\frac{1}{4}$ is approximately 68 (it's equal to $68 + \frac{1}{16}$), so the square root of 68 is approximately $8\frac{1}{4}$.

3. A 10-by-10 square has an area of 100 square units, so the square root of 106 is a little more than 100. To adjust the square to have area 106, we can append a 10-by-$\frac{6}{10}$ rectangle (which has an area of 6 square units), giving us a figure 10-by-$10\frac{6}{10}$ rectangle, which has the right area, but is not a square. Splitting the 10-by-$10\frac{6}{10}$ rectangle into two 10-by-$\frac{3}{10}$ rectangles and placing one of them on the bottom of the figure, we see that a $10\frac{3}{10}$-by-$10\frac{3}{10}$ square has an area that is approximately 106 square units (in fact, the picture shows us that $10\frac{3}{10} \times 10\frac{3}{10}$ equals $106 + \frac{9}{100}$). Therefore, the square root of 106 is approximately $10\frac{3}{10}$, or 10.3.

4. Following the reasoning of the last two problems, the figure below shows that the square root of 55 is approximately $7\frac{3}{7}$ [in fact, $(7 + \frac{3}{7})^2 = 55 + (\frac{3}{7})^2 = 55 + \frac{9}{49}$].

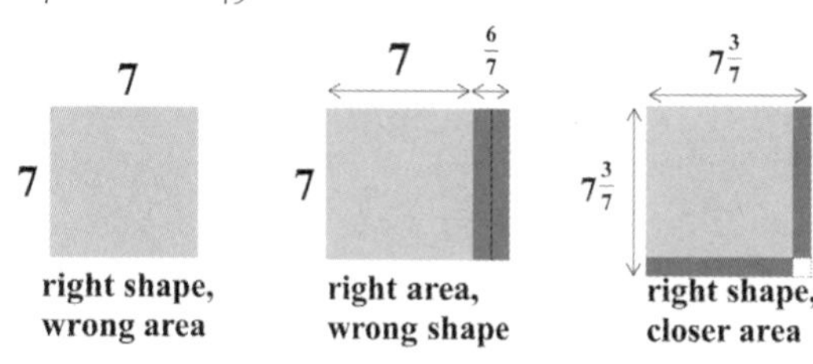


5. We want to find a square whose area is approximately $A^2 + b$ square units. Its side length will be approximately the square root of $A^2 + b$. Because $A^2 + b$ is a little more than $A^2$, we know that the square root of $A^2 + b$ is a little more than $A$. We start with an $A$-by-$A$ square, which has area $A^2$ square units, which is $b$ square units too small. That is, the $A$-by-$A$ square is the right shape (a square) but the wrong area. Because it is $b$ square units too small, we add $b$ square units on the right side of the square. To have area 4, the rectangle should have dimensions $A$ by $\frac{b}{A}$. Now, we have an $A$-by-$A + \frac{b}{A}$ rectangle, which has the correct area but the wrong shape. If we split the $A$-by-$\frac{b}{A}$ rectangle into two $A$-by-$\frac{b}{2A}$ rectangles, keeping one on the right of the square and placing the other at the bottom of the square, we have an $(A + \frac{b}{2A})$-by-$(A + \frac{b}{2A})$

square, except for the tiny $\frac{b}{2A}$-by-$\frac{b}{2A}$ square at the bottom right of the figure. This shows that $(A + \frac{b}{2A}) \times (A + \frac{b}{2A})$ is approximately $A^2 + b$ (it's equal to $A^2 + b + (\frac{b}{2A})^2 = A^2 + b + \frac{b^2}{4A^2}$), so the square root of $A^2 + b$ is approximately $A + \frac{b}{2A}$.

### Try This! A1-33

Start by noticing that 68 is a little more than 64, which is $8^2$, so we start with 8 as the whole number part of the solution.

$$\begin{array}{r} 8.\phantom{0} \\ \sqrt{68.} \end{array}$$

Subtracting 64 from 68 leaves a difference of 4, to which we append 00 and consider 400 in the next iteration of the method.

$$\begin{array}{r} 8.\phantom{0} \\ \sqrt{68.} \\ \underline{-64} \\ 400 \end{array}$$

Now comes the tricky part! We want to know what the tenths place of our square root is. To do this, we first double our estimate so far ($2 \times 8 = 16$) and fill in the blanks of the "equation":

$$\begin{array}{r} 16\_\_ \\ \underline{\times \ \_\_} \end{array}$$

with the *same* digit in both blanks so that the product is as large as possible but *still no larger than 400*. Because $\begin{array}{r} 162 \\ \underline{\times\ 2} \\ 324 \end{array}$ and $\begin{array}{r} 163 \\ \underline{\times\ 3} \\ 489 \end{array}$, so we choose 2 as our digit and place it in the tenths place of our solution and subtract 324 from 400:

$$\begin{array}{r} 8.2 \\ \sqrt{68.} \\ -64 \\ \underline{400} \\ \underline{-324} \\ 76 \end{array}$$

Now we append 00 to the end of 76 (giving us 7,600) and go through the above procedure.

We double 82 to get 164, and then we want to find the digit $d$ so that $\begin{array}{r} 164d \\ \underline{\times\ d} \end{array}$ is as large as possible, but still less than 7,600.

After a few tries, we see that 7,600 is between 1,644 × 4 = 6,576 and 1,645 × 5 = 8,225, so we place a 6 in the hundredths place.

Now, we have 8.24 as our approximation. We could stop there or continue. Subtracting 6,576 from 7,600 and appending a 00 at the right, we have 102,400.

$$\begin{array}{r} 8.24 \\ \sqrt{68.} \\ \underline{-64} \\ 400 \\ \underline{-324} \\ 7600 \\ \underline{-6576} \\ 102400 \end{array}$$

Next, we double 824. We get 1,648, so we want to find $d$ so that $1{,}648d \times d$ is as large as possible, but still less than 102,400. Because 16,486 × 6 = 98,916 and 16,487 × 7 = 115,409, we see that 6 goes in the thousandths place. The appendix shows continuing the process until we have an approximation of 8.24621 for $\sqrt{68}$.

## Equations

### What Kind of Graph Am I? A2-2

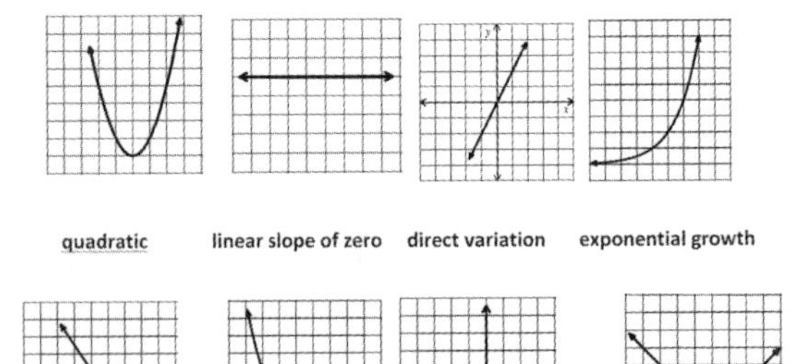

## Factor Lattices 2 A2-3

1. We see from the factor lattices of $xy^2$ and $xy^3$ that GCF $(xy^2, xy^3) \rightarrow xy^2$ and LCM $(xy^2, xy^3) \rightarrow xy^3$.

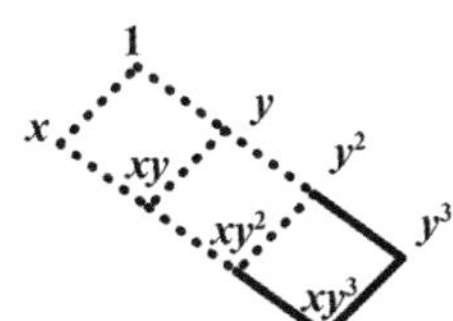


2. The factor lattice of $x^3y^4$ is shown with the factor lattices of $x^3y$ (dotted and dashed) and $x^2y^3$ (dotted and solid), with their overlap dotted. The GCF of $x^3y$ and $x^2y^3$ is $x^2y$; the LCM of $x^3y$ and $x^2y^3$ is $x^3y^3$.

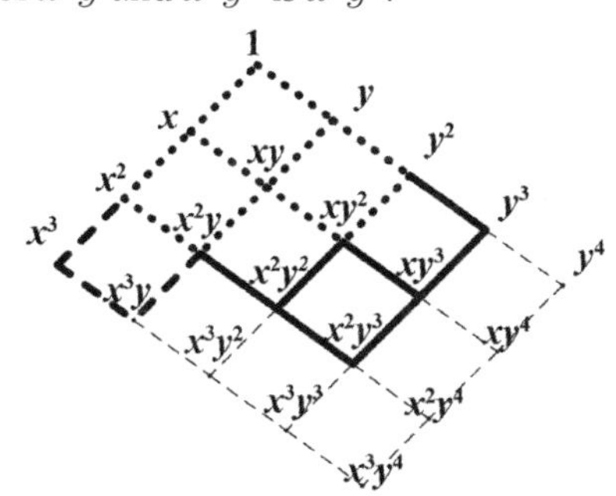


## Balance Beams A2-4

1. **a.** 11
   **b.** 20
   **c.** 20
   **d.** 4
   **e.** 4
   **f.** Answers will vary; accept any answer in which the value of the square is 2 less than the value of the triangle.
2. **a.** 4
   **b.** 0
   **c.** 3
   **d.** $\frac{4}{7}$
   **e.** No solution

## Can These Really Be True? A2-5

Answers will vary; sample answers are given.

1. 1 day + 1 week = 8 days
2. 1 twenty-dollar bill + 1 five-dollar bill = 25 dollars
3. 1 foot + 1 yard = 4 feet
4. 1 nickel + 1 quarter = 30 cents
5. 1 dollar + 1 quarter = 5 quarters
6. 1 nickel + 1 dime = 15 cents
7. Answers will vary.

## Let Me Count the Ways A2-8

1. There are sixty-six rectangles (of all dimensions from 1 by 1 to 1 by 11) in the figure.
   The expression for the number of rectangles in rectangle that is $n$ units long is $\frac{n(n+1)}{2}$.
2. There are ninety-one squares of all dimensions from 1 by 1 to 6 by 6. Within an $n$-by-$n$ square, there are a total of $n^2 + (n-1)^2 + (n-2)^2 + \ldots + 2^2 + 1^2 = \frac{n(n+1)(2n+1)}{6}$ squares of all possible dimensions.
3. $\frac{n(n+1)}{2}$

## Gauss Center Activity Cards A2-9

Counting Rectangles 1: The 2-by-3 rectangle contains eighteen rectangles of all dimensions.
Counting Rectangles 2: The 3-by-4 rectangle contains sixty rectangles of all dimensions.
Counting Rectangles 3: The 6-by-5 rectangle contains 315 rectangles of all dimensions.

## Mobile Madness A2-10

1. Star = 8, circle with line = 12, circle = 6
2. Heart = 18, circle = 9
3. Circle = 2.5, triangle = 5, diamond = 10, star = 7.5
4. Cylinder = 14, plus sign = 7, cube = 14

## Pattern Block Mobiles A2-11

1. You can add one rhombus or two equilateral triangles. If one triangle weighs 2 kilograms, the mobile weighs 48 kilograms.
2. To balance the mobile, place one triangle on the left string. Each triangle weighs 0.538 kilograms, rounded to the thousandths, or approximately 538 grams.

3. Place the hexagon and rhombus, one below the other, on the left bar. Place the trapezoid and triangle on the left string of the right bar.

## The Knight's Move Challenge A2-12

Because there are sixty-three squares that have to be landed on, there are at least sixty-three moves required. In fact, it is possible to hit every square exactly (if you're really careful and a little lucky), it's possible to land on all the squares in sixty-three moves.

## Something Is Fishy A2-14

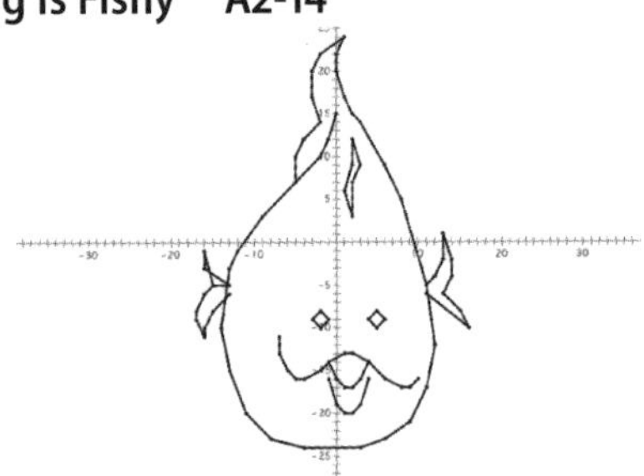

## Shape Equations A2-16

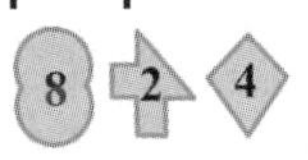

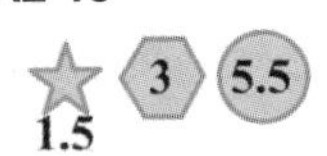

## Hypatia Center Activity Cards A2-17

Other answers are possible, but all answers must be explained.

**A.** **a.** 11, 3, 6, 10, 15, 21, 28, 36, 45, 55; $n$th term is $\frac{n(n+1)}{2}$, the $n$th triangular number.

**b.** 1, 2, 3, 5, 8, 13, 21, 34, 55, 89; these are the Fibonacci numbers, starting with the second 1.

**c.** 1, 8, 27, 64, 125, 216, 343; these are the consecutive cube numbers.

**d.** 1, 3, 7, 15, 31, 63, 127, 255; the $n$th term is $2^n - 1$. (The successive differences are increasing powers of 2.)

**e.** 1, 4, 9, 1, 6, 2, 5, 3, 6, 4, 9, 6, 4, 8, 1, 1 (If you start listing the square numbers and record the individual digits, this is what you get!)

**B.** **a.** 3, 7, 12, 18, 25, 33, 42, 52, 63; the successive differences start at 4 and increase by each term.

**b.** 0, 3, 0, 4, 1, 6, 3, 9, 6, 13, 10; pattern of "changes": plus 3, minus 3, plus 4, minus 3, plus 5 minus 3, plus 6, minus 3, plus 7, minus 3

**c.** 1, 3, 2, 4, 3, 5, 4, 6, 5, 7,6; pattern of "changes": plus 2, minus 1, plus 2 minus 1, …

**d.** 256, 128, 64, 32, 16, 8, 4, 2, 1, ½; each term is the previous term divided by 2.

**C.** Row 7 will have 13 faces, row 10 will have 19 faces and the $n$th row will have $2n - 1$ faces.

## Al-Khwarizmi Activity Center Cards A2-18

Answers are in the same row in the original (before being cut out and distributed).

**Teacher Note:** The observant student might note that some of these problems have restricted domains (in particular, only integer solutions "make sense"). Although the graphs and inequalities represent a continuum of possible solutions, only the integers within these solution sets are realistic. This is a fairly common convention in mathematics, but it is important to recognize that different contexts allow for different types of answers.

## Equations in All Forms A2-19

1. Slope = 3, equation: $y = 3x + 1$ (other forms of the equation are acceptable)

2. Slope = 2.5 ($y$-intercept = −6), equation: $y = 2.5x - 6$

3.

| *a* | *b* |
|---|---|
| 3 | 2 |
| -6 | 8 |
| -3 | 6 |
| 9 | -2 |
| 0 | 4 |
| 6 | 0 |

## Graphs and Equations A2-20

**1.** Slope = 0, equation: $y = 0x$

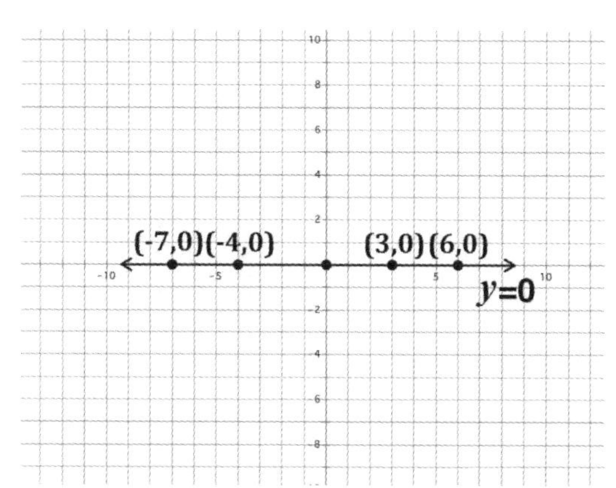


There will be a variety of different ordered pairs but may include: (0,4), (3, 2), (-3, 6), (-6, 8). Slope is $-\frac{2}{3}$.

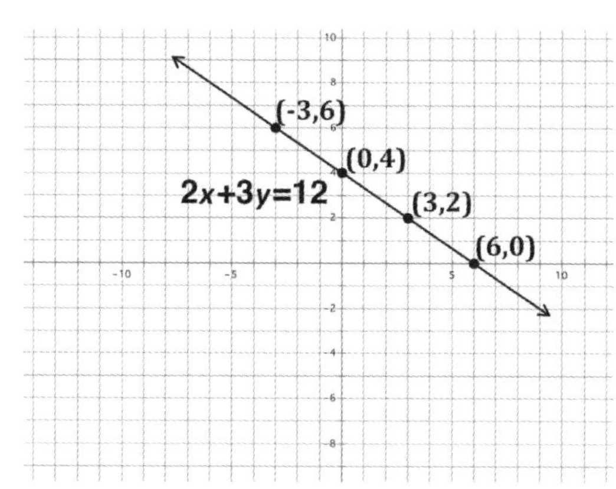


**2.** $x = 0$. The slope is undefined.

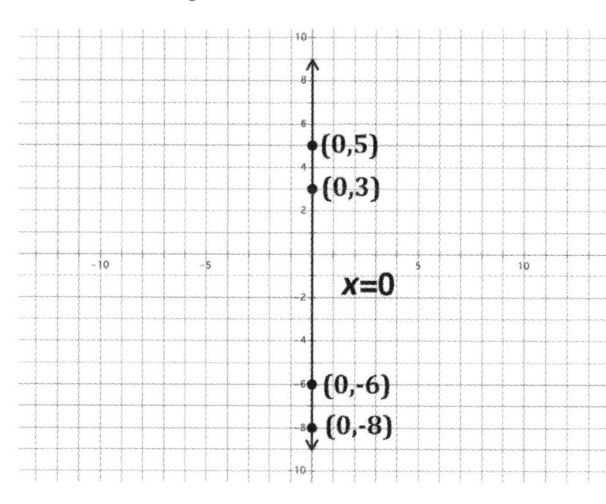


**3.** The slope is $\frac{1}{2}$.

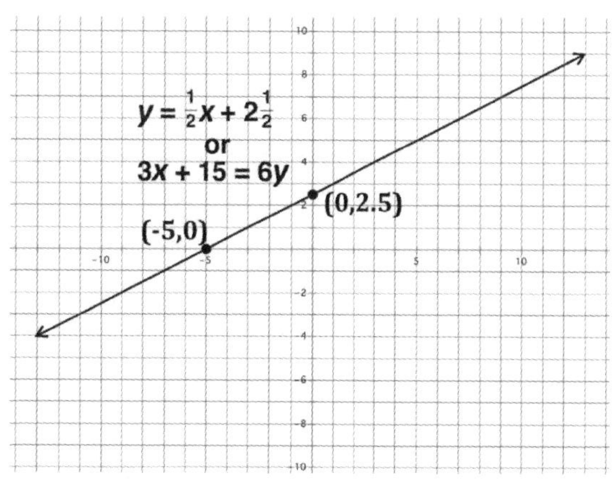


**4.** The slope is $-\frac{1}{2}$.

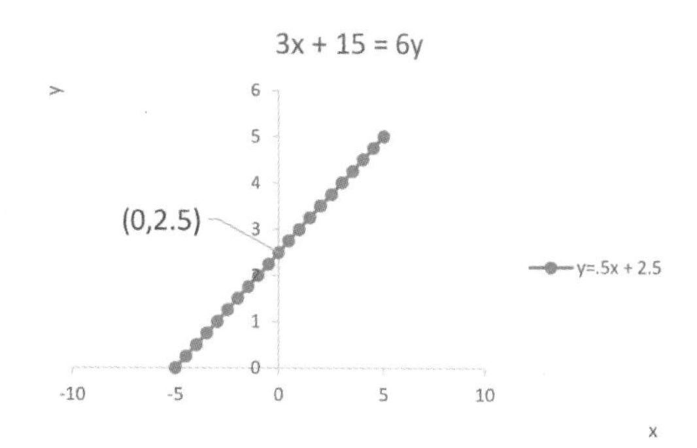


## Equation Puzzle A2-21

Pentagon = 11

Rhombus = 9

Circle = 6

The value of equation D is 20.

## Linear Systems A2-22

**1.** Graphing, substitution, elimination

**2.** Capris = 6 and dresses = 2

**3.** 5 hours is the break-even point.

**4.** 110° and 70°

## What Is Missing? A2-23

**1.** (4, 13), (0, 1), (1, 4), (4, 13), (0, 1), $(\frac{-1}{3}, 0)$. An equation for the line is $y = 3x + 1$; other equation forms are acceptable.

**2.** The slope is $-\frac{2}{3}$. The $y$-intercept is 4.

| $x$ | $y$ |
|---|---|
| 3 | 2 |
| -6 | 8 |
| -3 | 6 |
| 9 | -2 |
| 0 | 4 |
| 6 | 0 |

**3.** (3, -1), (0, 8), (–2, 24), (2, 0)

## Simultaneously! A2-25

**1.** Thirty-three tricycles and sixty-four bicycles

**2.** Twenty-seven tricycles and fifty-eight four-wheelers

**3.** 118 bicycles, 147 tricycles, and 100 unicycles

## Stephen and Cole A2-27

**1.** **a.** Stephen was traveling at 80 km/h.

**b.** 146 km between towns

**c.**

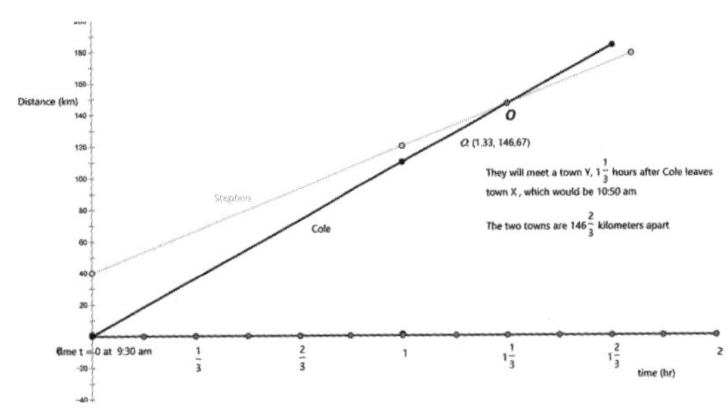


## Which Region Is It? A2-28

There are many solutions to the pizza problem. Any combination of $p$ pizzas and $b$ breadsticks satisfying the inequalities $p + b \geq 32$ and $10p + 4b \leq 200$. To spend exactly \$200 and buy exactly thirty-two items, Sophia needs to buy 12 pizzas and 20 breadsticks.

**1.** $p + b \geq 32$; $10p + 4b \leq 200$

**2.** **a.** $A = (12, 20)$, $B = (0, 50)$, $C = (0, 32)$, $D = (32, 0)$, $E = (20, 0)$

**b.** Here are the combinations: 12 pizzas and 20 breadsticks; 11 pizzas and 21 or 22 breadsticks; 10 pizzas and 22, 23, 24, or 25 breadsticks; 9 pizzas and 23–27 breadsticks; 8 pizzas and 24–30 breadsticks; 7 pizzas and 25–32 breadsticks; 6 pizzas and 26–35 breadsticks; 5 pizzas and 27–37 breadsticks; 4 pizzas and 28–40 breadsticks; 3 pizzas and 29–42 breadsticks; 2 pizzas and 30–45 breadsticks; 1 pizza and 31–47 breadsticks; 0 pizzas and 32–50 breadsticks.

**3.** Region I represents the possible orders.

## Solving Situations A2-30

**1.** **a.** $3.50m + 6.25w \leq 350$

**b.**

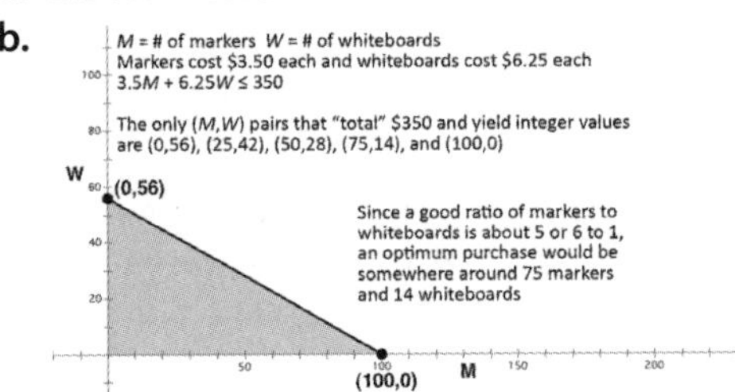


**c.** The possibilities are as follows:

| Markers | Whiteboards |
|---|---|
| 0 | 56 |
| 25 | 42 |
| 50 | 28 |
| 75 | 14 |
| 100 | 0 |

Seventy-five markers and 14 whiteboards are the optimal choices.

**2.** **a.** If $x$ represents the number of true-or-false questions and $y$ represents the number of multiple-choice questions, then $x + y = 20$ and $3x + 11y = 100$.

**b.**

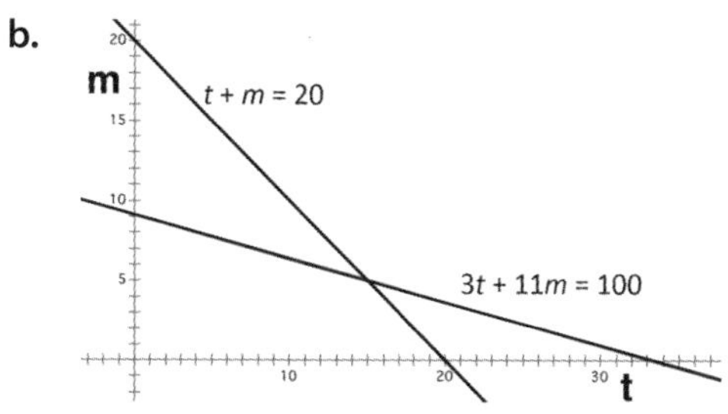


**c.** There are fifteen true-or-false questions.

**d.** There are five multiple-choice questions.

**3.** **a.** $25 \leq a \leq 40$ and $64 \leq t \leq 71$

**b.**

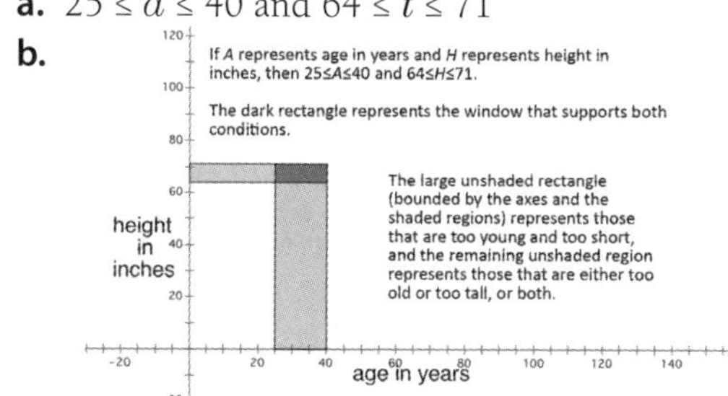


**4.** **a.** If $b$ represents the number of hours spent babysitting and $t$ represents the number of hours spent tutoring, then $5b + 10d \leq 90$ and $b + d \leq 12$.

**b.**

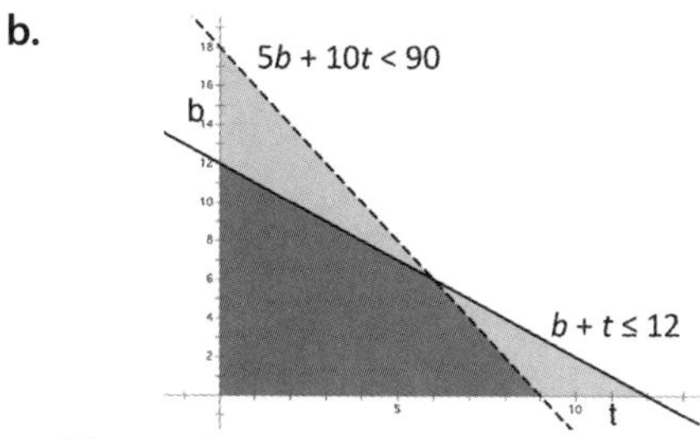


## Backpacking A2-32

**1.** Pictures will vary.

**2.** If $D$ is Nicole's distance from the base of the cliff (in meters) $t$ seconds after she started running, then $D = -8t + 43$.

**3.** At a speed of 8 meters per second, Nicole will reach the base of the cliff in $5\frac{3}{8}$ or 5.375 seconds.

## How Fast Did I Go? A2-33

**1.** The car will reach the exit in $\frac{3.6}{68}$ hours (approximately 0.05294 hours), which is about 190 seconds, or 3 minutes and 10 seconds. That means the highway patrol officer has 160 seconds (he lost 30 seconds waiting for his motorcycle to start) to go 3.6 miles to catch the speeding car. That will require him to drive a speed of 81 miles per hour.

$$\frac{3.6 \text{ miles}}{160 \text{ seconds}}, \times \frac{3{,}600 \text{ seconds}}{1 \text{ hour}} = \frac{3.6 \times 3{,}600 \text{ miles}}{160 \text{ hour}} = 81 \frac{\text{miles}}{\text{hour}}$$

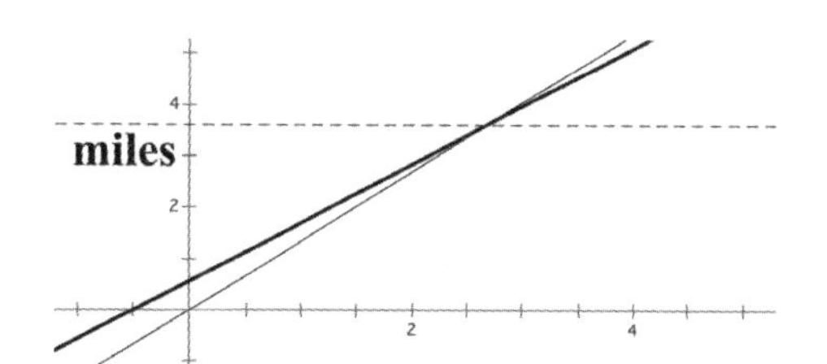


The thick line represents the distance the motorcyclist has traveled past the police officer (in miles), where time is measured in minutes after the police officer gets the motorcycle started. The $y$-intercept is $\frac{68}{120}$, because that is the distance the car would have traveled in 30 seconds at a speed of 68 miles per hour. The equation of the line is

$$y = \frac{68}{60}x + \frac{68}{120} \text{ or } y = \frac{17}{15}x + \frac{17}{30}$$

The horizontal line is the line $y = 3.6$, allowing us to visualize the time the car gets to the exit 3.6 miles away (by estimating/computing the $x$-coordinate of the point at which the thick line intersects the horizontal line). The coordinates of the point of intersection are $(\frac{91}{34}, 3.6)$, or approximately (2.67647, 3.6). The $y$-coordinate is computed by solving the equation

$$3.6 = \frac{17}{15}x + \frac{17}{30}$$

for $x$.

The thin line represents the distance traveled by the police officer to catch up with the car exactly 3.6 miles away from the starting point. The line will go through the origin (0,0) and the point $(\frac{91}{34}, 3.6)$, so its equation will be $y = \frac{612}{455}x$, because its $y$-intercept will be 0 and the slope will be $\frac{3.6}{91/34} = \frac{18}{5} \cdot \frac{34}{91} = \frac{612}{455}$.

2. Neither speed is reasonable and safe, because they are significantly above the speed limit.

3. Students should use their calculations in their explanations.

### Amanda's Subway Ride A2-34

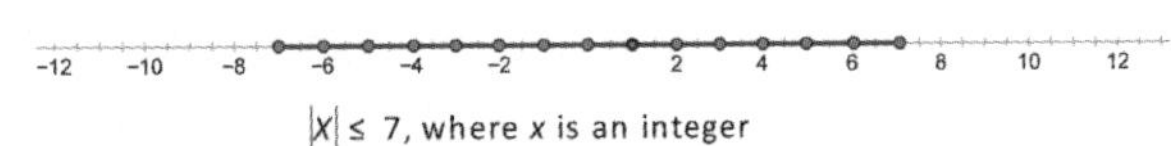


$|x| \le 7$, where $x$ is an integer

### My Number Line Representation A2-35

The matching representations are all in the same row in the original.

### Bottle Experiment Directions A2-36

Answers will vary but should be approximately quadratic. This is obvious in the graph. Emphasize that errors in measurement have an effect on data.

### Am I Linear, Quadratic, Exponential, or Other? A2-37

1. If these data represent a linear relationship, then the slope (change in output divided by change in input) will be constant. $\frac{38-3}{6-1} = \frac{35}{5} = 7$ but $\frac{11-3}{3-1} = \frac{8}{2} = 4$, so the relationship is not linear. Noticing that the outputs are all close to the squares of the inputs (two more, in fact), it's reasonable to guess that the relationship is quadratic given by the formula $n^2 + 2$.

2. Notice that $\frac{1}{9} = \frac{1^2}{3}$ and $-\frac{1}{27} = -\frac{1^3}{3} = (\frac{-1}{3})^3$. If $n$ is the input, then $(-\frac{1}{3})n$ is the output, so this is an exponential relationship.

3. The slope (change in output divided by change in input) is consistently 3, so this is a linear relationship with slope 3. This means that when the input is $n$, the output is $3n + b$. When the input is 2, the output is also 2, so $3(2) + b = 2$, which means $6 + b = 2$, so $b = -4$. When the input is $n$, the formula for the output is $3n - 4$.

4. When the input is $n$, the output is $|n|$, or the absolute value of $n$.

### Delilah and the Rooftop A2-38

1. 243 feet: $-16(2^2) + 128(2) + 51 = 243$

2. 6 seconds: ($-16t^2 + 128t + 51 = 243$ if $t = 2$ or 6)

3. 307 feet: Because outputs are equal at $t = 2$ and $t = 6$, the vertex of the parabola is at $t = 4$. The height at $t = 4$ is $-16(4^2) + 128(4) + 51 = 307$.

4. Approximately 8.38 seconds: Answers will vary because estimates are required. Although an exact solution requires the quadratic formula to solve when $-16t^2 + 128t + 51 = 0$, students can use a calculator (to graph the function and estimate the horizontal intercept) or use guess-and-check to narrow in on the intercept.

5. Approximately 1.41 seconds (on the way up) and 6.59 seconds (on the way down). Students can graph $y = -16x^2 + 128x + 51$ and $y = 200$ and then estimate the $x$ value of the intersection or use guess-and-check to find when the height is 200 feet.

6. 6.5 seconds

7. 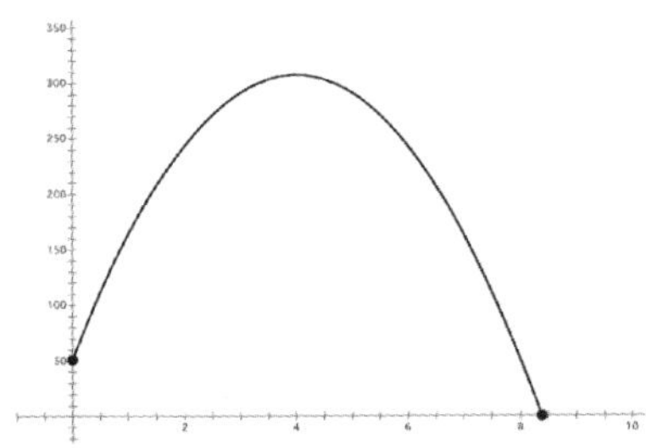

### Will She Catch It? A2-39

1. **a.** $h(50) = 5$
   **b.** $A = \frac{1}{25}$ or 0.04: If $-A(50)^2 + 105 = 5$, then $-A(50)^2 = -100$, so $-2{,}500A = -100$, so $A = -\frac{100}{-2{,}500} = \frac{1}{25}$, or 0.04.

2. $v = 20$ feet/second: Because $A = \frac{1}{25}$, $\frac{1}{25} = \frac{16}{v^2}$, so $v^2 = 16(25) = 400$, so $v = 20$ feet/second.

3. The velocity is approximately 13.64 miles per hour (a reasonable speed to throw a ball):

$$\frac{20 \text{ feet}}{1 \text{ second}} \times \frac{60 \text{ second}}{1 \text{ minute}} \times \frac{60 \text{ minute}}{1 \text{ hour}} \times \frac{1 \text{ mile}}{5280 \text{ feet}} =$$

$$\frac{20(60)(60) \text{ miles}}{5280 \text{ hour}} = \frac{72000}{5280 \text{ hour}} \approx 13.64 \frac{\text{miles}}{\text{hour}}$$

### Modeling Motion A2-40

**1.** Answers will vary. Some might guess that throwing the ball up into the air (rather than horizontally) will require more speed. Others might think it will go farther if you throw it at a 45° angle, so less speed is required.

**2.** $B = \frac{3}{50}$ or 0.06: For $h(50)$ to equal 5, we must have $-B(50)^2 + 50 + 105 = 5$, so $-2{,}500B = -150$; therefore, $B = \frac{-150}{-2{,}500} = \frac{3}{50}$, or 0.06.

**3.** $v$ is approximately 23.09 feet/second: Because $\frac{3}{50} = \frac{32}{v^2}$, $v^2 = \frac{32\,(50)}{3} = \frac{1{,}600}{3}$, which is approximately 533.33 feet/second, so $v$ is approximately 23.09 feet/second.

**4.** The velocity is approximately 15.74 mph, faster than before but still reasonable.

$$\frac{23.09 \text{ feet}}{1 \text{ second}} \times \frac{60 \text{ second}}{1 \text{ minute}} \times \frac{60 \text{ minute}}{1 \text{ hour}} \times \frac{1 \text{ mile}}{5280 \text{ feet}} =$$

$$\frac{23.09(60)(60) \text{ miles}}{5280 \text{ hour}} = \frac{83124 \text{ miles}}{5280 \text{ hour}} \approx \frac{15.74 \text{ miles}}{\text{hour}}$$

## Functions

### Working with Functions, Part 1 A3-2

**1.** $f(1) = 3(1) - 4 = 3 - 4 = -1$

**2.** $f(2) = 3(2) - 4 = 6 - 4 = 2$

**3.** $f(3) = 3(3) - 4 = 9 - 4 = 5$

**4.** $f(4) = 3(4) - 4 = 12 - 4 = 8$

**5.** $f(5) = 3(5) - 4 = 15 - 4 = 11$

**6.** No: $f(2 + 3) = f(5) = 11$ and $f(2) + f(3) = 2 + 5 = 7$

**7.** No: $f(3) - f(2) = 5 - 2 = 3$ and $f(3 - 2) = f(1) = -1$

**8.** $f(a + b) = 3(a + b) - 4 = 3a + 3b - 4$

**9.** It was shown that $f(2 + 3) \neq f(2) + f(3)$, so when $a = 2$ and $b = 3$, then $f(a + b)$ does not equal $f(a) + f(b)$.

**10.** No: because $f(a) + f(b) = 3a + 3b - 8$ and $f(a + b) = 3a + 3b - 4$; if they were *ever* equal, then there would be an $a$ and a $b$ so that $3a + 3b - 8 = 3a + 3b - 4$, but then -8 would have to equal -4, which is impossible.

**11.** No: If there was a number $a$ so that $f(-a) = -f(a)$, then for that value of $a$, $3(-a) - 4$ would equal $-(3a - 4)$. Simplifying both of these expressions and setting them equal to each other, it would be true that $-3a - 4 = -3a + 4$, but that would mean that $-4 = 4$, which is impossible.

**12.** Yes, but only if $a = 0$ (because, in that case $-a = a$). If such an $a$ existed, then $3(-a) - 4$ would equal $3a - 4$, so $0 = 6a$, which means that $a = 0$.

### Working with Functions, Part 2 A3-3

**1.**
**a.** $g(2) = 4 - 8 + 2 = -2$
**b.** $g(3) = 9 - 12 + 2 = -1$
**c.** $g(5) = 25 - 20 + 2 = 7$
**d.** $g(7) = 49 - 28 + 2 = 23$
**e.** $g(-2) = 4 + 8 + 2 = 14$
**f.** $g(-3) = 9 + 12 + 2 = 23$

**2.** Sample answer: Using the answers to item 1, we see that $g(3) = -1$, so $-g(3) = 1$ and $g(-3) = 23$. So $g(-a) \neq -g(a)$ when $a = 3$.

**3.** Yes, but only if $a = 0$. If $g(-a) = g(a)$, then $(-a)^2 - 4(-a) + 2 = a^2 - 4a + 2$, so $a^2 + 4a = a^2 - 4a$, so $8a = 0$; therefore, $a = 0$. Again, this makes sense because $0 = -0$, $f(-0) = f(0)$.

**4.** Because $g(2 + 3) = g(5) = 7$ and $g(2) + g(3) = -2 + -1 = -3$, $g(a + b)$ will not equal $g(a) + g(b)$ when $a = 2$ and $b = 3$.

**5.** Yes, there are many examples (for instance, if $a = b = 1$ or if $a = b = -1$ or if $a = 2$ and $b = \frac{1}{2}$). First, notice that $g(a + b) = (a + b)^2 - 4(a + b) + 2 = a^2 + 2ab + b^2 - 4a - 4b + 2$ and $g(a) + g(b) = a^2 - 4a + 2 + b^2 - 4b + 2 = a^2 + b^2 - 4a - 4b + 4$. If $g(a + b)$ ever equals $g(a) + g(b)$, then $a^2 + 2ab + b^2 - 4a - 4b + 2 = a^2 + b^2 - 4a - 4b + 4$, so $2ab = 2$, which can happen if $ab = 1$. For example, if $a = b = 1$, then $g(1 + 1) = g(2) = -2$ and $g(1) + g(1) = -1 + -1 = -2$. Also, $g(-1 + -1) = g(-2) = 14 = g(-1) + g(-1) = 1$ and $g(\frac{1}{2} + 2) = g(2.5) = -1.75 = g(\frac{1}{2}) + g(2)$.

6. No: If $g(-a) = -g(a)$, then $a^2 + 4a + 2 = -(a^2 - 4a + 2)$, so $a^2 + 4a + 2 = -a^2 + 4a - 2$, so $2a^2 = -4$, so $a^2 = -2$, which can't happen if a is a real number.

7. Yes, but only if $a = 0$ (again!): f $g(-a) = g(a)$, then $a^2 + 4a + 2 = a^2 - 4a + 2$, so $8a = 0$ and therefore $a = 0$.

8. $g(a - b) = (a - b)^2 - 4(a - b) + 2 = a^2 - 2ab + b^2 - 4a + 4b + 2 = a^2 + b^2 - 2ab - 4a + 4b + 2$

9. $g(a) - g(b) = a^2 - 4a + 2 - (b^2 - 4b + 2) = a^2 - 4a + 2 - b^2 + 4b - 2 = a^2 - b^2 - 4a + 4b$

## Am I a Function? A3-4

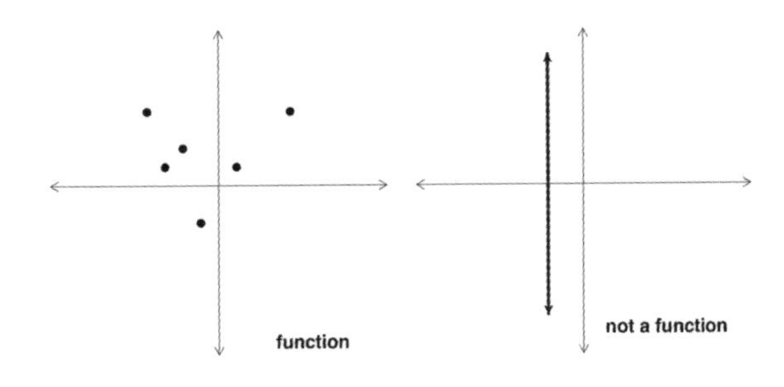


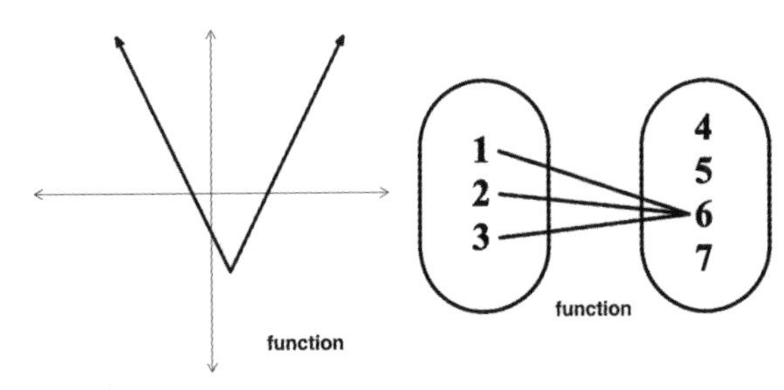


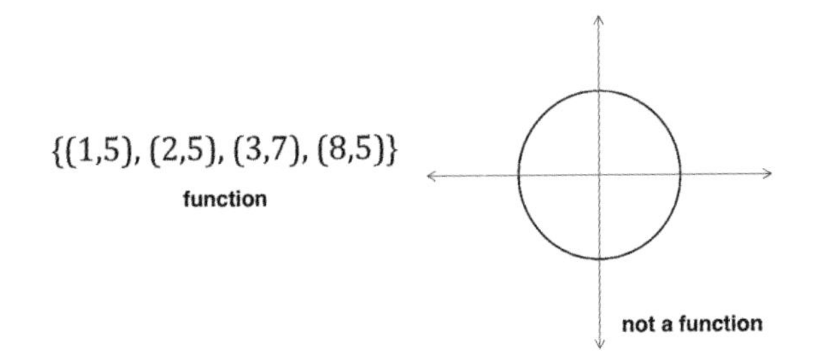


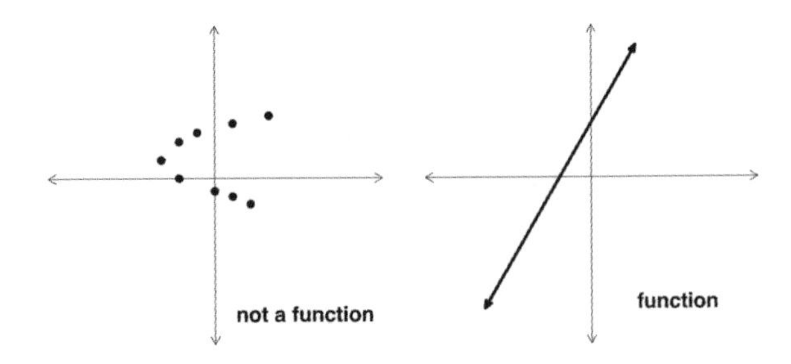


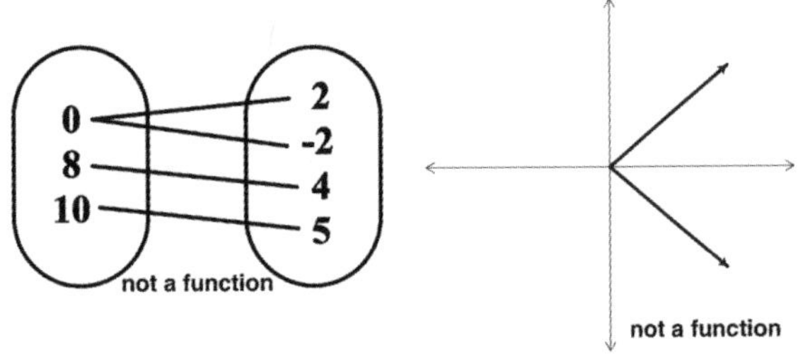


## Double-Wides A3-5

1. In each figure, each (horizontal) row has twice as many dots as each (vertical) column, so each figure is twice as "wide" as it is "high."

2. **a.** The fifth figure will have five rows of ten dots each, so $D(5) = 50$.

   **b.** The tenth figure will have ten rows of twenty dots each, so $D(10) = 200$.

   **c.** The one hundredth figure will have 100 rows, each having 200 dots, so $D(100) = 20{,}000$.

3. Figure $n$ will have $n$ rows each having $2n$ dots, for a total of $n \times 2n = 2n^2$ dots, so $D(n) = 2n^2$.

4. $D(2.5)$ should equal $2 \times (2.5)^2 = 2 \times 6.25 = 12.5$.
   $D(-3.7)$ should equal $2 \times (-3.7)^2 = 2 \times 13.69 = 27.38$.
   We know that $D(2.5)$ and $D(-3.7)$ are not double-wide numbers, because figure numbers need to be counting numbers. Furthermore, although we might be able to make sense of 2½ rows of five dots (because ½ row of five dots could contain 2.5 dots), we can't have a negative number of rows or a negative number of dots, so there's no way to think of $D(-3.7)$ as a double-wide number.

5. The number 128 will be a double-wide number if we can find a counting number n so that $D(n) = 128$. That is, $2n^2 = 128$, so $n^2 = 64$, and therefore $n = 8$ or $-8$. Because 8 is a counting number, 128 is a double-wide number (in fact, 128 is the eighth double-wide number). This also shows that $D(8)$ and $D(-8)$ both equal 128.

6. Because $D(x) = 2x^2$, we need to find an $x$ so that $2x^2 = 75$, which means that $x^2 = 37.5$, so $x = \sqrt{37.5}$ or $-\sqrt{37.5}$. Because neither of these numbers are counting numbers, 75 is not a double-wide number.

7. The $n$th triple-wide figure should have $n$ rows each having $3n$ dots, for a total of $3n^2$ dots, so the $n$th triple-wide number is $3n^2$. Therefore, the first four triple-wide numbers are 3 × 12 = 3, 3 × 22 = 12, 3 × 32 = 27, and 3 × 42 = 48.

## Square Numbers A3-6

1. The fifth square number will be represented by a square array with five rows having five dots each, for a total of 5 × 5 = 25 dots, and the fifteenth square number will be represented by the square array having fifteen rows of fifteen dots each, for a total of 15 × 15 = 225 dots. Therefore, the fifth square number is 25 and the fifteenth square number is 225.

2. Because the $n$th square number is represented by a square having $n$ rows with $n$ dots each, the $n$th square number is $n \times n = n^2$, and therefore $S(n) = n^2$.

3. $S(1.5) = (1.5)^2 = 2.25$, $S(-3) = (-3)^2 = 9$, and $S(\sqrt{7}) = (\sqrt{7})^2 = 7$.

4. $S(5.5) = (5.5)^2 = 30.25$, because $\sqrt{30.25} = 5.5$, so $x$ could be 5.5 (or –5.5).

5. Because $S(-5.5) = (-5.5)^2$ also equals 30.25, *another* $x$ could be -5.5.

6. Although it might seem so, a little extra thought will convince you that although every *positive* output has two corresponding inputs (if $N$ is the output, then $\sqrt{N}$ and $-\sqrt{N}$ will both be inputs), if $N = 0$ is the output, the only possible input is 0 (after all, 0 = -0, right?).

## Translate This A3-7

**a.** The height of the instruments 0 seconds after release is 90,000 feet, *or* when the instruments are released, their height is 90,000 feet. This is a true statement.

**b.** The instruments are higher 7 seconds after being released than they are 15 seconds after being released. This statement is true, because the instruments are falling toward the ocean.

**c.** The height of the instruments 15 seconds after being dropped is 95,000 feet. This is not true, because they were dropped from a height of 90,000 feet.

**d.** The height of the instruments after 15 seconds is 0 feet, or the instruments hit the ocean 15 seconds after being released. This is not true, because it will take the instruments much longer to hit the ocean.

**e.** The instruments hit the ocean 90,000 seconds after being released. Because 90,000 seconds is 25 hours, that seems like an awfully long time. It is not true.

**f.** The height of the instruments is the same after 14 and 15 seconds. This is not true, because falling instruments would not remain at the same height for one second.

7. **a.** $f(60)$. Remember that the units of $t$ are seconds, so 1 minute needs to be converted to 60 seconds.

**b.** $\frac{f(5) - f(0)}{5 - 0}$

8. **a.** $f(15) = 90{,}000 - 16(15)^2 = 86{,}400$ feet
**b.** $f(60) = 32{,}400$
**c.** In 5 seconds, the instruments fall 400 feet, so the average velocity is $\frac{400}{5}$, or 80 feet per second.
**d.** If $90{,}000 - 16x^2 = 0$, then $16x^2 = 90{,}000$, so $x^2 = \frac{90{,}000}{16} = 5{,}625$, so $x = 75$ seconds, or 1 minute and 15 seconds.

## Translate That A3-9

**1.** **a.** $f(150) = \$200$; $f(200) = 200 + 0.20(200 - 180) = \$204$

**b.** a is correct: $f(180) = 200$, but $f(200) = 200 + 0.20(200 - 180) = \$204$.

**2.** The cost is \$200 as long as you drive no more than 180 miles. For every mile beyond 180 miles that you drive, you will be charged an additional 20 cents.

**3.** 430 miles: Rachel certainly drove more than 180 miles, because she paid more than \$200. If $f(x)$ equals 250, then $200 + 0.20(x - 180) = 250$, so $0.2(x - 180) = 50$, so $x - 180 = \frac{50}{0.2} = 250$, so $x = 430$ miles.

**4.** **a.** $f(m) = m$

**b.** $m = 205$ miles: If $f(m) = m$, then $200 + 0.20(m - 180) = m$, so $200 + 0.2m - 36 = m$. This means $164 = 0.8m$; therefore, $m = \frac{164}{0.8} = 205$ miles.

## How Cold Is It? A3-10

**1.** $F(15) = 1.8(15) + 32 = 59$

**2.** **a.**

| C | F(C) = 1.8C + 32 | T(C) = 2C + 30 |
|---|---|---|
| 5 | 41 | 40 |
| 10 | 50 | 50 |
| 15 | 59 | 60 |
| 20 | 68 | 70 |

**b.**

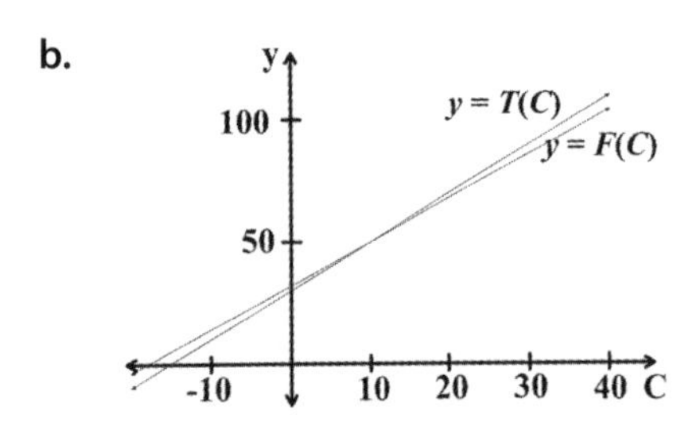


**c.** 50° is the only common temperature.

**d.** $5 \le |T(C) - F(C)| \le 5$ if $-15 \le C \le 35$ (one can solve algebraically or graphically)

## Absolute Translations A3-11

**1.** The number is seven; equation: $5n = 35$

**2.** The number is either four or negative four; equation: $n^2 = 16$

**3.** The number could be two or eight; equation: $|n - 5| = 3$ or $|5 - n| = 3$

**4.** Same as c.

## Mix and Match A3-12

**1.** a – C – III

**2.** b – D – IV

**3.** c – A – I

**4.** d – B – II

## Which Exponential Function Is It? A3-14

**1.** Because $\frac{f(2)}{f(1)} = \frac{f(3)}{f(2)} = 6$, this function could be exponential.

**2.** **a.** $f(1) = Ab$ and $f(2) = Ab^2$

**b.** $A = 2$ and $b = 6$: Using $f(1) = 12$ and $f(1) = Ab$, we have $Ab = 12$. Using $f(2) = 72$ and $f(2) = Ab^2$, we have $Ab^2 = 432$. We need to solve the two equations $Ab = 12$ and $Ab^2 = 72$ for $A$ and $b$. Substituting $A = \frac{12}{b}$ (from the first equation) into the second equation, we have $(= \frac{12}{b})\, b^2 = 72$, so $12b = 72$, so $b = 6$ (the common ratio from the table). Substituting $b = 6$ into the equation $A = \frac{12}{b}$, we get $A = 2$.

**c.** Using the formula $f(x) = 2(6^x)$, $f(3) = 2(6^x) = 2(216) = 432$, which matches the information from the table.

**3.** $g(x) = 0.75(2x)$ or $\frac{3}{4}2^x$: using the first two inputs given, $Ab^2 = 3$ and $Ab^4 = 12$. Therefore, $\frac{12}{3} = \frac{(Ab^4)}{(Ab^2)}$, so $4 = b^2$ and therefore $b = 2$ or $-2$. If $b = 2$, then (substituting into the equation $Ab^2 = 3$), we get $4A = 3$, so $A = \frac{3}{4}$, or 0.75, giving us the equation $g(x) = 0.75(2x)$. Notice, though, that if $b = -2$, then we solve for $Ab^2 = 3$, we still get $A = 0.75$, but then $f(7) = 0.75(-2)7 = -96$, not 96. Therefore, $b$ must equal 2 and $g(x) = 0.75(2x)$.

### Coupon Stacks A3-15

1. **a.** $2^{10} = 1{,}024$ coupons
   **b.** $2^{26} = 67{,}108{,}864$ coupons

2. **a.**

$$67{,}108{,}864 \text{ coupons} \times \frac{1 \text{ inch}}{200 \text{ coupons}} \times \frac{1 \text{ foot}}{12 \text{ inches}} \times \frac{1 \text{ mile}}{5{,}280 \text{ feet}} = \frac{67{,}108{,}864}{200 \times 12 \times 5{,}280} \text{ miles}$$

which is approximately 5.3 miles!

**b.**

$$67{,}108{,}864 \text{ coupons} \times \frac{5 \text{ pounds}}{3{,}000 \text{ coupons}} \times \frac{1 \text{ ton}}{2{,}000 \text{ pounds}} = \frac{67{,}108{,}864 \times 5}{3{,}000 \times 2{,}000} \text{ pounds}$$

which is approximately 55.9 tons!

### Tea with Mrs. Wiley A3-16

1. $f(t) = 137 \times 2^{-kt} + 75$: We have $R = 75$ and $A + R = 212$, $A$ must equal $212 - 75 = 137°$.

2. Using the formula for $f(t)$ and $k = 0.245$, $f(10) = 137(2^{2.45}) + 75$, which is approximately 100.07.

3. 8:59:30, which is 30 seconds before 9:00 a.m.: We need to find $t$ so that $f(t) = 85$, so $137(2^{245t}) + 75 = 85$. We can graph the equation $y = 137(2^{245x}) + 75$ on a graphing calculator and trace to find a point with a $y$-value of 85 *or* graph the equations $y = 137(2^{0.245x}) + 75$ and $y = 85$ and locate the intersection point. (Note that students will likely *not* be familiar with logarithms at this point, which is another strategy that could be used.) In any case, the value of $t$ is approximately 15.4 minutes (or 15.5 minutes rounded to the nearest half-minute), so Steve needs to pour the boiling water 15.5 minutes before Mrs. Wiley and Jackie arrive at 9:15. That is, at 30 seconds before 9:00 (officially, 8:59:50 a.m.).

### Piecing Together the Cost A3-18

1.

| Weight (in pounds) | 0.75 | 1.283 | 5.001 | 6.999 | 8.7 |
|---|---|---|---|---|---|
| **Cost (in dollars)** | **2.56** | **3.02** | **4.86** | **5.32** | **6.20** |

2. The cost of mailing a book via the U.S. Postal Service is \$2.56 for packages weighing up to 1 pound. Every additional pound (or fraction of a pound) up to a total weight of 7 pounds costs 46 cents. Beyond a total of 7 pounds, every additional full (or fraction of) a pound costs 44 cents.

3. **a.** Alpha Wireless service costs \$40 for up to 500 minutes of use plus an additional 5 cents for every minute over 500 minutes.
   Beta Cellular service costs \$25 for up to 300 minutes of use plus an additional 10 cents for every minute over 300 minutes.

   **b.** Answers will vary because Beta is better for 425 minutes, but Alpha is better for 475 minutes (and they cost the same for 450 minutes). Alpha Wireless will cost Sophia \$40 per month for 425–475 minutes of use. $B(425) = 25 + 0.1(125) = \$37.50$ and $B(475) = 25 + 0.1(175) = \$42.50$. It's interesting that $B(450) = \$40$, so the middle of her minutes range is the time at which both services charge \$40.

### Finite Differences Revisited 1 A3-20

1. The outputs consistently increase by four when the inputs increase by one, so it's reasonable to guess that $f(6) = 19 + 4 = 23$.

2. The function appears to be linear. If we plotted the points, they would lie on a line of slope 4. Its equation, in slope-intercept form, is $y = 4x - 1$, but there are many equivalent forms.

3.

| $x$ | f(x) | Difference: f(x + 1) – f(x) |
|---|---|---|
| 1 | $M + B$ | $M$ |
| 2 | $2M + B$ | $M$ |
| 3 | $2M + B$ | $M$ |
| 4 | $4M + B$ | $M$ |
| 5 | $5M + B$ | $M$ |

**4.** Table on the left: $f(x) = 6x - 3$; table on the right: $f(x) = -4x + 13$

| $x$ | f(x) | f(x + 1) – f(x) |
|---|---|---|
| 1 | 3 | 6 |
| 2 | 9 | 6 |
| 3 | 15 | 6 |
| 4 | 21 | 6 |

| $x$ | f(x) | f(x + 1) – f(x) |
|---|---|---|
| 1 | 9 | -4 |
| 2 | 5 | -4 |
| 3 | 1 | -4 |
| 4 | -3 | -4 |

## Finite Differences Revisited 2 A3-21

**1.**

| $x$ | g(x) | Diff (x): g(x + 1) – g(x) | Diff (x + 1) – Diff (x) |
|---|---|---|---|
| 1 | $A + B + C$ | $(4A + 2B + C) - (A + B + C) = 3A + B$ | $(5A + B) - (3A + B) = 2A$ |
| 2 | $4A + 2B + C$ | $(9A + 3B + C) - (4A + 2B + C) = 5A + B$ | $(7A + B) - (5A + B) = 2A$ |
| 3 | $9A + 6B + C$ | $(16A + 4B + C) - (9A + 3B + C) = 7A + B$ | $(9A + B) - (7A + B) = 2A$ |
| 4 | $16A + 4B + C$ | $(25A + 5B + C) - (16A + 4B + C) = 9A + B$ | $(11A + B) - (9A + B) = 2A$ |
| 5 | $25S + 5B + C$ | $(36A + 6B + C) - (25A + 5B + C) = 11A + B$ | $(13A + B) - (11A + B) = 2A$ |
| 6 | $36A + 6B + C$ | $(49A + 7B + C) - (36A + 6B + C) = 13A + B$ | $(15A + B) - (13A + B) = 2A$ |

**2.**

| $x$ | g(x) | Diff (x): g(x + 1) – g(x) | Diff (x + 1) – Diff (x) |
|---|---|---|---|
| 0 | $0 + 0 + C = C$ | $(A + B + C) - C = A + B$ | $(3A + B) - (A + B) = 2A$ |

**3.** DIFF($x$) is not a constant, but DIFF($x$ + 1) – DIFF($x$) is a constant, because it always equals $2A$.

**4.**

| $x$ | g(x) | Diff (x): g(x + 1) – g(x) | Diff (x + 1) – Diff (x) |
|---|---|---|---|
| 1 | $1\ (=A + B + C)$ | $2 = (3A + B)$ | $4\ (=2A)$ |
| 2 | 3 | $6 = (5A + B)$ | $4\ (=2A)$ |
| 3 | 9 | $10 = (7A + B)$ | $4\ (=2A)$ |
| 4 | 19 | $14 = (9A + B)$ | $4\ (=2A)$ |
| 5 | 33 | $18 = (11A + B)$ | $4\ (=2A)$ |
| 6 | 51 | $22 = (13A + B)$ | $4\ (=2A)$ |

Using the table from problem 1, we saw that if the function is quadratic, DIFF($x$ + 1) – DIFF($x$) will be equal to $2A$. That is, $2A = 4$ for our function, so $A = 2$. Now, looking at the DIFF($x$) column when $x = 1$, because $2 = 3A + B$ and $A = 2$, we know that $2 = 3(2) + B$, so $2 = 6 + B$; therefore, $B = -4$. Finally, comparing the two tables in the $g(x)$ column, $g(1) = 1$ in our table and $g(1) = A + B + C$ in the problem 1 table, so $1 = A + B + C$. Because $A = 2$ and $B = -4$, we have $1 = 2 - 4 + C$, so $1 = -2 + C$; therefore, $C = 3$. The function $g(x) = 2x^2 - 4x + 3$ fits the information in the table!

**5.**

| $x$ | g(x) | Diff (x): g(x + 1) – g(x) | Diff (x + 1) – Diff (x) |
|---|---|---|---|
| 1 | $3 = A + B + C$ | $1 = 3A + B$ | $2 = 2A$ |
| 2 | 4 | 3 | 2 |
| 3 | 7 | 5 | 2 |
| 4 | 12 | 7 | 2 |
| 5 | 19 | 9 | 2 |
| 6 | 28 | | |

Because the second differences [DIFF($x$ + 1) – DIFF($x$)] are always 2, it's reasonable to try to fit a quadratic function to the data. According to Table 1, this means $2A = 2$, so $A = 1$. Because $g(2) - g(1)$ equals 1 according to our table and should equal $3A + B$ by problem 1, we have $1 = 3A + B = 3(1) + B = 3 + B$, so $B = -2$. Finally, $g(1) = 3$ in our problem and should equal $A + B$

$+ C$, we have $3 = 1 - 2 + C = -1 + C$, so $C = 4$. Checking the table, we see that $g(x) = x^2 - 2x + 4$ fits.

**6.** What about this one?

| $x$ | $f(x)$ |
|---|---|
| 1 | 1 |
| 2 | 4 |
| 3 | 9 |
| 4 | 16 |
| 5 | 25 |
| 6 | 36 |

**a.** $f(x) = x^2$ would be a much less complicated function matching the information in the table.

**b.** Because we were told that $f(x) = x^2 + (x - 1)(x - 2)(x - 3)(x - 4)(x - 5)(x - 6)$, and we know that $(x - 1)(x - 2)(x - 3)(x - 4)(x - 5)(x - 6)$ will equal 0 whenever $x$ is equal to 1, 2, 3, 4, 5, or 6, we see that $f(x)$ will equal $x^2 + 0$ whenever $x$ is equal to 1, 2, 3, 4, 5, or 6.

**c.** According to the alternative formula $f(7) = 7^2 + (6)(5)(4)(3)(2)(1) = 49 + 720$, which equals 769.